INTRODUCTORY DISCRETE MATHEMATICS

INTRODUCTORY DISCRETE MATHEMATICS

Gordon Vining
Joel Helms

Kruger Brentt
Publishers
2025

Kruger Brentt Publishers UK. LTD.
Company Number 9728962
Regd. Office: 68 St Margarets Road, Edgware, Middlesex HA8 9UU

ISBN: 978-1-78715-376-9

For information on all our publications visit our website at http://krugerbrentt.com/

PREFACE

"Introductory Discrete Mathematics" embarks on an illuminating journey into the foundational principles and techniques of discrete mathematics, offering readers a comprehensive understanding of the fundamental concepts, methods, and applications in the field. This book serves as an essential resource for students, educators, and enthusiasts seeking to explore the rich landscape of discrete mathematics and its diverse range of real-world applications.

Discrete mathematics provides a rigorous framework for studying mathematical structures and objects that are distinct and separate, rather than continuous. From exploring sets, relations, and functions to analyzing graphs, trees, and combinatorial structures, discrete mathematics plays a crucial role in computer science, cryptography, operations research, and other fields where discrete structures are prevalent.

The primary objective of this book is to provide readers with a comprehensive introduction to discrete mathematics, covering key topics such as set theory, logic, proof techniques, relations and functions, graph theory, and combinatorics. Each chapter offers insights, theoretical foundations, practical examples, and problem-solving strategies to help readers develop a deep understanding of discrete mathematical concepts and methods.

Contributions from leading experts in the field explore various aspects of discrete mathematics, including techniques for modeling discrete structures, algorithms for solving combinatorial problems, and applications in diverse domains such as computer networking, cryptography, and data analysis. Special emphasis is placed on the interdisciplinary nature of discrete mathematics, with discussions on its connections to computer science, engineering, and other branches of mathematics.

As the digital revolution continues to transform society and drive innovation in technology, the importance of discrete mathematics in solving complex problems and optimizing algorithms has never been greater. By mastering the principles and techniques of discrete mathematics, individuals can gain valuable insights into the underlying structures of discrete systems, develop efficient algorithms, and tackle a wide range of practical problems with precision and creativity.

We extend our sincere gratitude to the contributors whose expertise and dedication have enriched this book. It is our hope that "Introductory Discrete Mathematics" will serve as a valuable resource and inspiration for readers interested in exploring the beauty and utility of discrete mathematics and its applications in various fields of science, technology, and engineering.

Gordon Vining

Joel Helms

CONTENTS

CHAPTER-1

INTRODUCTION TO DISCRETE MATHEMATICS

INTRODUCTION: AN OVERVIEW

Discrete mathematics serves as the bedrock of theoretical computer science and modern mathematics, offering a structured approach to understanding and solving problems that deal with distinct, countable elements. Unlike continuous mathematics, which focuses on concepts like calculus and real numbers, discrete mathematics examines discrete structures such as integers, graphs, and statements in logic. Its applications span various fields including computer algorithms, cryptography, and data analysis, making it indispensable in both theoretical and practical domains of computer science and beyond. By exploring topics like set theory, combinatorics, and graph theory, discrete mathematics provides the foundational tools for tackling complex problems with precision and clarity.Top of Form Bottom of Form

1.1 DISCRETE MATHEMATICS

dis·crete / dis'krët.

Adjective: Individually separate and distinct.

Synonyms: separate - detached - distinct - abstract.

Defining *discrete mathematics* is hard because defining *mathematics* is hard. What is mathematics? The study of numbers? In part, but you also study functions and lines and triangles and parallelepipeds and vectors and .... Or perhaps you want to say that mathematics is a collection of tools that allow you to solve problems. What sort of problems? Okay, those that involve numbers, functions, lines, triangles, ... Whatever your conception of what mathematics is, try applying the concept of "discrete" to it, as defined above. Some math fundamentally deals with *stuff* that is individually separate and distinct.

In an algebra or calculus class, you might have found a particular set of numbers (maybe the set of numbers in the range of a function). You would represent this set as an interval: $[0,\infty)$ is the range of $f(x) = x^2$ since the set of outputs of the function are all real numbers 0 and greater. This set of numbers is NOT discrete. The numbers in the set are not separated by much at all. In fact, take any two numbers in the set and there are infinitely many more between them which are also in the set.

Discrete math could still ask about the range of a function, but the set would not be an interval. Consider the function which gives the number of children of each person reading this. What is the range? I'm guessing it is something like $\{0, 1, 2, 3\}$. Maybe 4 is in there too. But certainly there is nobody reading this that has 1.32419 children. This output set is discrete because the elements are separate. The inputs to the function also form a discrete set because each input is an individual person.

One way to get a feel for the subject is to consider the types of problems you solve in discrete math. Here are a few simple examples:

Investigate!

Note: *Throughout the text you will see* ***Investigate!*** activities like this one. Answer the questions in these as best you can to give yourself a feel for what is coming next.

The most popular mathematician in the world is throwing a party for all of his friends. As a way to kick things off, they decide that everyone should shake hands. Assuming all 10 people at the party each shake hands with every other person (but not themselves, obviously) exactly once, how many handshakes take place?

At the warm-up event for Oscar's All Star Hot Dog Eating Contest, Al ate one hot dog. Bob then showed him up by eating three hot dogs. Not to be outdone, Carl ate five. This continued with each contestant eating two more hot dogs than the previous contestant. How many hot dogs did Zeno (the 26th and final contestant) eat? How many hot dogs were eaten all together?

After excavating for weeks, you finally arrive at the burial chamber. The room is empty except for two large chests. On each is carved a message (strangely in English):

If this chest is empty, then the other chest's message is true.	This chest is filled with treasure or the other chest contains deadly scorpions.

You know exactly one of these messages is true. What should you do?

Back in the days of yore, five small towns decided they wanted to build roads directly connecting each pair of towns. While the towns had plenty of money to build roads as long and as winding as they wished, it was very important that the roads not intersect with each other (as stop signs had not yet been invented). Also, tunnels and bridges were not allowed. Is it possible for each of these towns to build a road to each of the four other towns without creating any intersections?

Complete the activity mentioned above before moving forward

One reason it is difficult to define discrete math is that it is a very broad description which encapsulates a large number of subjects. In this course we will study four main topics: combinatorics (the theory of ways things combines; in particular, how to count these ways), sequences, symbolic logic, and graph theory. However, there are other topics that belong under the discrete umbrella, including computer science, abstract algebra, number theory, game theory, probability, and geometry (some of these, particularly the last two, have both discrete and non-discrete variants).

Ultimately the best way to learn what discrete math is about is to do it. Let's get started! Before we can begin answering more complicated (and fun) problems, we must lay down some foundation. We start by reviewing mathematical statements, sets, and functions in the framework of discrete mathematics.

1.2 MATHEMATICAL STATEMENTS

Investigate!

While walking through a fictional forest, you encounter three trolls guarding a bridge. Each is either a *knight,* who always tells the truth, or a *knave,* who always lies. The trolls will not let you pass until you correctly identify each as either a knight or a knave. Each troll makes a single statement:

Troll 1: If I am a knave, then there are exactly two knights here.

Troll 2: Troll 1 is lying.

Troll 3: Either we are all knaves or at least one of us is a knight.

Which troll is which?

Attempt the above activity before proceeding

In order to do mathematics, we must be able to talk and write about mathematics. Perhaps your experience with mathematics so far has mostly involved finding answers to problems. As we embark towards more advanced and abstract mathematics, writing will play a more prominent role in the mathematical process.

Communication in mathematics requires more precision than many other subjects, and thus we should take a few pages here to consider the basic building blocks: *mathematical statements.*

Atomic and Molecular Statements

A statement is any declarative sentence which is either true or false. A statement is atomic if it cannot be divided into smaller statements, otherwise it is called molecular.

Example 1.2.1

These are statements (in fact, atomic statements):

Telephone numbers in the USA have 10 digits.

The moon is made of cheese.

42 is a perfect square.

Every even number greater than 2 can be expressed as the sum of two primes.

$$3 + 7 = 12$$

And these are not statements:

Would you like some cake?

The sum of two squares.

$$1 + 3 + 5 + 7 + \cdots + 2n + 1.$$

Go to your room!

$$3 + x = 12$$

The reason the sentence "$3 + x = 12$" is not a statement is that it contains a variable. Depending on what x is, the sentence is either true or false, but right now it is neither. One way to make the *sentence* into a *statement* is to specify the value of the variable in some way. This could be done by specifying a specific substitution, for example, "$3 + x = 12$ where $x = 9$," which is a true statement. Or you could capture the free variable by *quantifying* over it, as in, "for all values of x, $3 + x = 12$," which is false.

We will discuss quantifiers in more detail at the end of this section.

You can build more complicated (molecular) statements out of simpler (atomic or molecular) ones using logical connectives. For example, this is a molecular statement:

Telephone numbers in the USA have 10 digits and 42 is a perfect square.

Note that we can break this down into two smaller statements. The two shorter statements are *connected* by an "and." We will consider 5 connectives: "and" (Sam is a man and Chris is a woman), "or" (Sam is a man or Chris is a woman), "if. . . , then. . . " (if Sam is a man, then Chris is a woman), "if and only if" (Sam is a man if and only if Chris is a woman), and "not" (Sam is not a man). The first four are called binary connectives (because they connect two statements) while "not" is an example of a unary connective (since it applies to a single statement).

These molecular statements are of course still statements, so they must be either true or false. The absolutely key observation here is that which truth value the molecular statement achieves is completely determined by the type of connective and the truth values of the parts. We do not need to know what the parts actually say, only whether those parts are true or false. So to analyse logical connectives, it is enough to consider propositional variables (sometimes called sentential variables), usually capital letters in the middle of the alphabet: P, Q, R, S ,...... We think of these as standing in for (usually atomic) statements, but there are only two *values* the variables can achieve: true or false.[1] We also have symbols for the logical connectives: $\wedge$, $\vee$, $\rightarrow$, $\leftrightarrow$, $\neg$.

Logical Connectives.

$P \wedge Q$ is read "P and Q," and called a **conjunction.**

$P \vee Q$ is read "P or Q," and called a **disjunction.**

$P \rightarrow Q$ is read "if P then Q," and called an **implication** or **conditional.**

$P \leftrightarrow Q$ is read "P if and only if Q," and called a **biconditional.**

$\neg P$ is read "not P," and called a **negation.**

The truth value of a statement is determined by the truth value(s) of its part(s), depending on the connectives:

Truth Conditions for Connectives.

$P \wedge Q$ is true when both P and Q are true.

$P \vee Q$ is true when P or Q or both are true.

$P \rightarrow Q$ is true when P is false or Q is true or both.

$P \leftrightarrow Q$ is true when P and Q are both true, or both false.

$\neg P$ is true when P is false.

Note that for us, or is the **inclusive or** (and not the sometimes used *exclusive* or) meaning that $P \vee Q$ is in fact true when both P and Q are true. As for the other connectives, "and" behaves as you would expect, as does negation. The biconditional (if and only if) might seem a little strange, but you should think of this as saying the two parts of the statements are *equivalent* in that they have the same truth value. This leaves only the conditional P Q which has a slightly different meaning in

[1]In computer programming, we should call such variables Boolean variables

mathematics than it does in ordinary usage. However, implications are so common and useful in mathematics, that we must develop fluency with their use, and as such, they deserve their own subsection.

Implications

An implication or conditional is a molecular statement of the form

$$P \to Q$$

where P and Q are statements. We say that

P is the hypothesis (or **antecedent**).

Q is the conclusion (or **consequent**).

An implication is true provided P is false or Q is true (or both), and false otherwise. In particular, the only way for $P \to Q$ to be false is for P to be true and Q to be false.

Easily the most common type of statement in mathematics is the implication. Even statements that do not at first look like they have this form conceal an implication at their heart. Consider the Pythagorean Theorem. Many a college freshman would quote this theorem as "$a^2 + b^2 = c^2$." This is absolutely not correct. For one thing, that is not a statement since it has three variables in it. Perhaps they imply that this should be true for any values of the variables? So $1^2 + 5^2 = 2^2$??? How can we fix this? Well, the equation is true as long as a and b are the legs of a right triangle and c is the hypotenuse. In other words:

- If a and b are the legs of a right triangle with hypotenuse c, then $a^2 + b^2 = c^2$.

This is a reasonable way to think about implications: our claim is that the conclusion ("then" part) is true, but on the assumption that the hypothesis ("if" part) is true. We make no claim about the conclusion in situations when the hypothesis is false.[2]

Still, it is important to remember that an implication is a statement, and therefore is either true or false. The truth value of the implication is determined by the truth values of its two parts. To agree with the usage above, we say that an implication is true either when the hypothesis is false, or when the conclusion is true. This leaves only one way for an implication to be false: when the hypothesis is true and the conclusion is false.

Example 1.2.2

Consider the statement:

If Bob gets a 90 on the final, then Bob will pass the class.

[2]However, note that in the case of the Pythagorean Theorem, it is also the case that if $a_2 + b_2\ c_2$, then a and b are the legs of a right triangle with hypotenuse c. So we could have also expressed this theorem as a biconditional: "a and b are the legs of a right triangle with hypotenuse c if and only if $a_2 + b_2\ c_2$"

This is definitely an implication: P is the statement "Bob gets a 90 on the final," and Q is the statement "Bob will pass the class."

Suppose I made that statement to Bob. In what circumstances would it be fair to call me a liar? What if Bob really did get a 90 on the final, and he did pass the class? Then I have not lied; my statement is true. However, if Bob did get a 90 on the final and did not pass the class, then I lied, making the statement false. The tricky case is this: what if Bob did not get a 90 on the final? Maybe he passes the class, maybe he doesn't. Did I lie in either case? I think not. In these last two cases, P was false, and the statement So $P \rightarrow Q$ is true when either P is false or Q is true.

Just to be clear, although we sometimes read P Q as "P *implies* Q", we are not insisting that there is some causal relationship between the statements P and Q. In particular, if you claim that $P \rightarrow Q$ is false, you are not saying that P does not imply Q, but rather that P is true and Q is false.

Example 1.2.3

Decide which of the following statements are true and which are false. Briefly explain.

- If 1 = 1, then most horses have 4 legs.
- If 0 = 1, then 1 = 1.
- If 8 is a prime number, then the 7624th digit of π is an 8.
- If the 7624th digit of π is an 8, then 2 + 2 = 4.

Solution

All four of the statements are true. Remember, the only way for an implication to be false is for the if part to be true and the then part to be false.

Here both the hypothesis and the conclusion are true, so the implication is true. It does not matter that there is no meaningful connection between the true mathematical fact and the fact about horses.

Here the hypothesis is false and the conclusion is true, so the implication is true.

I have no idea what the 7624th digit of π is, but this does not matter. Since the hypothesis is false, the implication is automatically true.

Similarly here, regardless of the truth value of the hypothesis, the conclusion is true, making the implication true.

It is important to understand the conditions under which an implication is true not only to decide whether a mathematical statement is true, but in order to prove that it is. Proofs might seem scary (especially if you have had a bad high school geometry experience) but all we are really doing is explaining (very carefully) why a statement is true. If you understand the truth conditions for an implication, you already have the outline for a proof.

Direct Proofs of Implications

To prove an implication $P \to Q$, it is enough to assume P, and from it, deduce Q.

Perhaps a better way to say this is that to prove a statement of the form $P \to Q$ directly, you must explain why Q is true, but you get to assume P is true first. After all, you only care about whether Q is true in the case that P is as well.

There are other techniques to prove statements (implications and others) that we will encounter throughout our studies, and new proof techniques are discovered all the time. Direct proof is the easiest and most elegant style of proof and has the advantage that such a proof often does a great job of explaining why the statement is true.

Example 1.2.4

Prove: If two numbers a and b are even, then their sum $a + b$ is even.

Solution

Proof. Suppose the numbers a and b are even. This means that $a = 2k$ and $b = 2j$ for some integers k and j. The sum is then $a + b = 2k + 2j = 2(k + j)$. Since $k + j$ is an integer, this means that $a + b$ is even.

Notice that since we get to assume the hypothesis of the implication, we immediately have a place to start. The proof proceeds essentially by repeatedly asking and answering, "what does that mean?" Eventually, we conclude that it means the conclusion.

This sort of argument shows up outside of math as well. If you ever found yourself starting an argument with "hypothetically, let's assume . . . ," then you have attempted a direct proof of your desired conclusion.

An implication is a way of expressing a relationship between two statements. It is often interesting to ask whether there are other relationships between the statements. Here we introduce some common language to address this question.

Converse and Contrapositive

The **converse** of an implication $P \to Q$ is the implication $Q \to P$. The converse is NOT logically equivalent to the original implication. That is, whether the converse of an implication is true is independent of the truth of the implication.

The **contrapositive** of an implication $P \to Q$ is the statement $\neg Q \to \neg P$. An implication and its contrapositive are logically equivalent (they are either both true or both false).

Mathematics is overflowing with examples of true implications which have a false converse. If a number greater than 2 is prime, then that number is odd. However, just because a number is odd does not mean it is prime. If a shape is a square, then it is a rectangle. But it is false that if a shape is a rectangle, then it is a square.

However, sometimes the converse of a true statement is also true. For example, the Pythagorean theorem has a true converse: if $a^2 + b^2 = c^2$, then the triangle with sides a, b, and c is a right triangle. Whenever you encounter an implication in mathematics, it is always reasonable to ask whether the converse is true.

The contrapositive, on the other hand, always has the same truth value as its original implication. This can be very helpful in deciding whether an implication is true: often it is easier to analyze the contrapositive.

Example 1.2.5

True or false: If you draw any nine playing cards from a regular deck, then you will have at least three cards all of the same suit. Is the converse true?

Solution

True. The original implication is a little hard to analyze because there are so many different combinations of nine cards. But consider the contrapositive: If you don't have at least three cards all of the same suit, then you don't have nine cards. It is easy to see why this is true: you can at most have two cards of each of the four suits, for a total of eight cards (or fewer).

The converse: If you have at least three cards all of the same suit, then you have nine cards. This is false. You could have three spades and nothing else. Note that to demonstrate that the converse (an implication) is false, we provided an example where the hypothesis is true (you do have three cards of the same suit), but where the conclusion is false (you do not have nine cards).

Understanding converses and contrapositives can help understand implications and their truth values:

Example 1.2.6

Suppose I tell Sue that if she gets a 93% on her final, then she will get an A in the class. Assuming that what I said is true, what can you conclude in the following cases:

- Sue gets a 93% on her final.
- Sue gets an A in the class.
- Sue does not get a 93% on her final.
- Sue does not get an A in the class.

Solution

Note first that whenever $P \to Q$ and P are both true statements, Q must be true as well. For this problem, take P to mean "Sue gets a 93% on her final" and Q to mean "Sue will get an A in the class."

We have $P \to Q$ and P, so Q follows. Sue gets an A.

You cannot conclude anything. Sue could have gotten the A because she did extra credit for example. Notice that we do not know that if Sue gets an A, then she gets a 93% on her final. That is the converse of the original implication, so it might or might not be true.

The contrapositive of the converse of $P \to Q$ is $\neg P \to \neg Q$, which states that if Sue does not get a 93% on the final, then she will not get an A in the class. But this does not follow from the original implication. Again, we can conclude nothing. Sue could have done extra credit.

What would happen if Sue does not get an A but did get a 93% on the final? Then P would be true and Q would be false. This makes the implication $P \to Q$ false! It must be that Sue did not get a 93% on the final. Notice now we have the implication $\neg Q \to \neg P$ which is the contrapositive of $P \to Q$. Since $P \to Q$ is assumed to be true, we know $\neg Q \to \neg P$ is true as well.

As we said above, an implication is not logically equivalent to its converse, but it is possible that both the implication and its converse are true. In this case, when both $P \to Q$ and $Q \to P$ are true, we say that P and Q are equivalent and write $P \to Q$. This is the biconditional we mentioned earlier.

If and only if

$P \to Q$ is logically equivalent to $(P \to Q) \wedge (Q \to P)$.

Example: Given an integer n, it is true that n is even if and only if n^2 is even. That is, if n is even, then n^2 is even, as well as the converse: if n^2 is even, then n is even.

You can think of "if and only if" statements as having two parts: an implication and its converse. We might say one is the "if" part, and the other is the "only if" part. We also sometimes say that "if and only if" statements have two directions: a forward direction $(P \to Q)$ and a backwards direction ($P \to Q$, which is really just sloppy notation for $Q \to P$).

Let's think a little about which part is which. Is $P \to Q$ the "if" part or the "only if" part? Consider an example.

Example 1.2.7

Suppose it is true that I sing if and only if I'm in the shower. We know this means both that if I sing, then I'm in the shower, and also the converse, that if I'm in the shower, then I sing. Let P be the statement "if I sing, then I'm in the shower." Which part of the if and only if statement is this?

What we are really asking for is the meaning of "I sing if I'm in the shower" and "I sing *only if* I'm in the shower." When is the first one (the "if" part) *false*? When I am in the shower but not singing. That is the same condition on being false as the statement "if I'm in statement, "I sing," and Q be, "I'm in the shower." So $P \to Q$ is the hand, to say, "I sing only if I'm in the shower" is equivalent to saying "if I sing, then I'm in the shower," so the "only if" part is $P \to Q$.

It is not terribly important to know which part is the "if" or "only if" part, but this does illustrate something very, very important: *there are many ways to state an implication!*

Example 1.2.8

Rephrase the implication, "if I dream, then I am asleep" in as many different ways as possible. Then do the same for the converse.

Solution

The following are all equivalent to the original implication:

- I am asleep if I dream.
- I dream only if I am asleep.
- In order to dream, I must be asleep.
- To dream, it is necessary that I am asleep.
- To be asleep, it is sufficient to dream.
- I am not dreaming unless I am asleep.

The following are equivalent to the converse (if I am asleep, then I dream):

- I dream if I am asleep.
- I am asleep only if I dream.
- It is necessary that I dream in order to be asleep.
- It is sufficient that I be asleep in order to dream.
- If I don't dream, then I'm not asleep.

Hopefully you agree with the above example. We include the "necessary and sufficient" versions because those are common when discussing mathematics. In fact, let's agree once and for all what they mean.

Necessary and Sufficient

- "P is necessary for Q" means $Q \rightarrow P$.
- "P is sufficient for Q" means $P \rightarrow Q$.
- If P is necessary and sufficient for Q, then $P \rightarrow Q$.

To be honest, I have trouble with these if I'm not very careful. I find it helps to keep a standard example for reference.

Example 1.2.9

Recall from calculus, if a function is differentiable at a point c, then it is continuous at c, but that the converse of this statement is not true (for example, $f(x)=|x|$ at the point 0). Restate this fact using "necessary and sufficient" language.

Solution. It is true that in order for a function to be differentiable at a point c, it is necessary for the function to be continuous at c. However, it is not necessary that a function be differentiable at c for it to be continuous at c.

It is true that to be continuous at a point c, it is sufficient that the function be differentiable at c. However, it is not the case that being continuous at c is sufficient for a function to be differentiable at c.

Thinking about the necessity and sufficiency of conditions can also help when writing proofs and justifying conclusions. If you want to establish some mathematical fact, it is helpful to think what other facts would *be enough* (be sufficient) to prove your fact. If you have an assumption, think about what must also be necessary if that hypothesis is true.

Predicates and Quantifiers

Investigate!

Consider the statements below. Decide whether any are equivalent to each other, or whether any imply any others.

- You can fool some people all of the time.
- You can fool everyone some of the time.
- You can always fool some people.
- Sometimes you can fool everyone.

Attempt the above activity before proceeding

It would be nice to use variables in our mathematical sentences. For example, suppose we wanted to claim that if n is prime, then $n + 7$ is not prime. This looks like an implication. I would like to write something like

$$P(n) \rightarrow \neg P(n+7)$$

where $P(n)$ means "n is prime." But this is not quite right. For one thing, because this sentence has a **free variable** (that is, a variable that we have not specified anything about), it is not a statement. A sentence that contains variables is called a **predicate.**

Now, if we plug in a specific value for n, we do get a statement. In fact, it turns out that no matter what value we plug in for n, we get a true implication in this case. What we really want to say is that *for all* values of n, if n is prime, then $n + 7$ is not. We need to *quantify* the variable.

Although there are many types of *quantifiers* in English (e.g., many, few, most, etc.) in mathematics we, for the most part, stick to two: existential and universal.

Universal and Existential Quantifiers

The existential quantifier is $\exists$ and is read "there exists" or "there is."

For example, asserts that there is a number less than 0.

$$\exists x(x < 0)$$

The universal quantifier is $\forall$ and is read "for all" or "every." For example, asserts that every number is greater than or equal to 0.

$$\forall x(x \geq 0)$$

As with all mathematical statements, we would like to decide whether quantified statements are true or false. Consider the statement

$$\forall x \exists y(y < x).$$

You would read this, "for every x there is some y such that y is less than x." Is this true? The answer depends on what our *domain of discourse* is: when we say "for all" x, do we mean all positive integers or all real numbers or all elements of some other set? Usually this information is implied. In discrete mathematics, we almost always quantify over the *natural numbers,* 0, 1, 2, . . . , so let's take that for our domain of discourse here.

For the statement to be true, we need it to be the case that no matter what natural number we select, there is always some natural number that is strictly smaller. Perhaps we could let y be x-1? But here is the problem: what if $x = 0$? Then $y = 1$ and that is *not a number!* (in our domain of discourse). Thus we see that the statement is false because there is a number which is less than or equal to all other numbers. In symbols,

$\exists x \forall y(y \geq x)$.

To show that the original statement is false, we proved that the negation was true. Notice how the negation and original statement compare. This is typical.

Quantifiers and Negation

$\neg\forall xP(x)$ is equivalent to $\exists x\neg P(x)$.

$\neg\forall xP(x)$ is equivalent to $\exists x\neg P(x)$.

Essentially, we can pass the negation symbol over a quantifier, but that causes the quantifier to switch type. This should not be surprising: if not everything has a property, then something doesn't have that property. And if there is not something with a property, then everything doesn't have that property.

Implicit Quantifiers.

It is always a good idea to be precise in mathematics. Sometimes though, we can relax a little bit, as long as we all agree on a convention. An example of such a convention is to assume that sentences containing predicates with free variables are intended as statements, where the variables are universally quantified.

For example, do you believe that if a shape is a square, then it is a rectangle? But how can that be true if it is not a statement? To be a little more precise, we have two predicates: $S(x)$ standing for "x is a square" and $R(x)$ standing for "x is a rectangle". The sentence we are looking at is,

$$S(x) \to R(x).$$

This is neither true nor false, as it is not a statement. But come on! We all know that we meant to consider the statement, and this is what our convention tells us to consider.

$$\forall x(S(x) \to R(x)),$$

Similarly, we will often be a bit sloppy about the distinction between a predicate and a statement. For example, we might write, let $P(n)$ be the statement, "*n* is prime," which is technically incorrect. It is implicit that we mean that we are defining P n to be a predicate, which for each n becomes the statement, *n* is prime.

Exercises

For each sentence below, decide whether it is an atomic statement, a molecular statement, or not a statement at all.

- Customers must wear shoes.
- The customers wore shoes.
- The customers wore shoes and they wore socks.

Classify each of the sentences below as an atomic statement, a molecular statement, or not a statement at all. If the statement is molecular, say what kind it is (conjunction, disjunction, conditional, biconditional, negation).

- The sum of the first 100 odd positive integers.

- Everybody needs somebody sometime.
- The Broncos will win the Super Bowl or I'll eat my hat.
- We can have donuts for dinner, but only if it rains.
- Every natural number greater than 1 is either prime or composite.
- This sentence is false.

Suppose P and Q are the statements:

P: Jack passed math. **Q:** Jill passed math.

- Translate "Jack and Jill both passed math" into symbols.
- Translate "If Jack passed math, then Jill did not" into symbols.
- Translate "$P \vee Q$" into English.
- Translate "$\neg(P \wedge Q) \to Q$" into English.

Suppose you know that if Jack passed math, then so did Jill. What can you conclude if you know that:

Jill passed math?

Jill did not pass math?

Determine whether each molecular statement below is true or false, or whether it is impossible to determine. Assume you do not know what my favorite number is (but you do know that 13 is prime).

- If 13 is prime, then 13 is my favorite number.
- If 13 is my favorite number, then 13 is prime.
- If 13 is not prime, then 13 is my favorite number.
- 13 is my favorite number or 13 is prime.
- 13 is my favorite number and 13 is prime.
- 7 is my favorite number and 13 is not prime.
- 13 is my favorite number or 13 is not my favorite number.

In my safe is a sheet of paper with two shapes drawn on it in colored crayon. One is a square, and the other is a triangle. Each shape is drawn in a single color. Suppose you believe me when I tell you that *if the square is blue, then the triangle is green.* What do you therefore know about the truth value of the following statements?

- The square and the triangle are both blue.
- The square and the triangle are both green.
- If the triangle is not green, then the square is not blue.

- If the triangle is green, then the square is blue.
- The square is not blue or the triangle is green.

Again, suppose the statement "if the square is blue, then the triangle is green" is true. This time however, assume the converse is false. Classify each statement below as true or false (if possible).

- The square is blue if and only if the triangle is green.
- The square is blue if and only if the triangle is not green.
- The square is blue.
- The triangle is green.

Consider the statement, "If you will give me a cow, then I will give you magic beans." Decide whether each statement below is the converse, the contrapositive, or neither.

- If you will give me a cow, then I will not give you magic beans.
- If I will not give you magic beans, then you will not give me a cow.
- If I will give you magic beans, then you will give me a cow.
- If you will not give me a cow, then I will not give you magic beans.
- You will give me a cow and I will not give you magic beans.
- If I will give you magic beans, then you will not give me a cow.

Consider the statement "If Oscar eats Chinese food, then he drinks milk."

Write the converse of the statement.

Write the contrapositive of the statement.

Is it possible for the contrapositive to be false? If it was, what would that tell you?

Suppose the original statement is true, and that Oscar drinks milk. Can you conclude anything (about his eating Chinese food)?

Explain

Suppose the original statement is true, and that Oscar does not drink milk. Can you conclude anything (about his eating Chinese food)? Explain.

You have discovered an old paper on graph theory that discusses the *viscosity* of a graph (which for all you know, is something completely made up by the author). A theorem in the paper claims that "if a graph satisfies *condition* (V), then the graph is *viscous*." Which of the following are equivalent ways of stating this claim? Which are equivalent to the *converse* of the claim?

- A graph is viscous only if it satisfies condition (V).
- A graph is viscous if it satisfies condition (V).
- For a graph to be viscous, it is necessary that it satisfies condition (V).
- For a graph to be viscous, it is sufficient for it to satisfy condition (V).
- Satisfying condition (V) is a sufficient condition for a graph to be viscous.
- Satisfying condition (V) is a necessary condition for a graph to be viscous.
- Every viscous graph satisfies condition (V).
- Only viscous graphs satisfy condition (V).

Write each of the following statements in the form, "if . . . , then . . . " Careful, some of the statements might be false (which is alright for the purposes of this question).

- To lose weight, you must exercise.
- To lose weight, all you need to do is exercise.
- Every American is patriotic.
- You are patriotic only if you are American.
- The set of rational numbers is a subset of the real numbers.
- A number is prime if it is not even.
- Either the Broncos will win the Super Bowl, or they won't play in the Super Bowl.

Which of the following statements are equivalent to the implication, "if you win the lottery, then you will be rich," and which are equivalent to the converse of the implication?

- Either you win the lottery or else you are not rich.
- Either you don't win the lottery or else you are rich.
- You will win the lottery and be rich.
- You will be rich if you win the lottery.
- You will win the lottery if you are rich.
- It is necessary for you to win the lottery to be rich.
- It is sufficient to win the lottery to be rich.
- You will be rich only if you win the lottery.
- Unless you win the lottery, you won't be rich.
- If you are rich, you must have won the lottery.

- If you are not rich, then you did not win the lottery.
- You will win the lottery if and only if you are rich.
- Let $P(x)$ be the predicate, "$3x + 1$ is even."
- Is $P(5)$ true or false?
- What, if anything, can you conclude about $\exists xP(x)$ from the truth value of $P(5)$?
- What, if anything, can you conclude about $\forall xP(x)$ from the truth value of $P(5)$?
- Let $P(x)$ be the predicate, "$4x + 1$ is even."
- Is $P(5)$ true or false?

What, if anything, can you conclude about $\exists xP(x)$ from the truth value of $P(5)$?

What, if anything, can you conclude about $\forall xP(x)$ from the truth value of $P(5)$?

For a given predicate $P(x)$, you might believe that the statements $\forall xP(x)$ or $\exists xP(x)$ are either true or false. How would you decide if you were correct in each case? You have four choices: you could give an example of an element n in the domain for which $P(n)$ is true or for which $P(n)$ if false, or you could argue that no matter what n is, $P(n)$ is true or is false.

1. What would you need to do to prove $\forall xP(x)$ is true?
2. What would you need to do to prove $\forall xP(x)$ is false?
3. What would you need to do to prove $\exists xP(x)$ is true?
4. What would you need to do to prove $\exists xP(x)$ is false?

Suppose $P(x, y)$ is some binary predicate defined on a very small domain of discourse: just the integers 1, 2, 3, and 4. For each of the 16 pairs of these numbers, $P(x, y)$ is either true or false, according to the following table (x values are rows, y values are columns).

	1	2	3	4
1	T	F	F	F
2	F	T	T	F
3	T	T	T	T
4	F	F	F	F

For example, $P(1,3)$ is false, as indicated by the F in the first row, third column.

Use the table to decide whether the following statements are true or false.

$$\forall x\exists yP(x, y).$$

$$\forall y\exists xP(x, y).$$

$$\exists x \forall y P(x, y).$$

$$\exists y \forall x P(x, y).$$

Translate into symbols. Use $E(x)$ for "x is even" and $O(x)$ for "x is odd."

1. No number is both even and odd.
2. One more than any even number is an odd number.
3. There is prime number that is even.
4. Between any two numbers there is a third number.
5. There is no number between a number and one more than that number.

Translate into English:

$$\forall x(E(x) \rightarrow E(x + 2)).$$

$$\forall x \exists y(\sin(x) = y).$$

$$\forall y \exists x(\sin(x) = y).$$

$$\forall x \forall y(x^3 = y^3 \rightarrow x = y).$$

Suppose $P(x)$ is some predicate for which the statement $\forall x P(x)$ is true. Is it also the case that $\exists x P(x)$ is true? In other words, is the statement $\forall x P(x) \rightarrow \exists x P(x)$ always true? Is the converse always true? Assume the domain of discourse is non-empty.

For each of the statements below, give a domain of discourse for which the statement is true, and a domain for which the statement is false.

(a) $\forall x \exists y\ (y^2 = x)$.

(b) $\forall x \forall y\ (x < y \rightarrow \exists z(x < z < y))$.

(c) $\exists x \forall y \forall z\ (y < z \rightarrow y \leq x \leq z)$.

Consider the statement, "For all natural numbers *n*, if n is prime, then n is solitary." You do not need to know what *solitary* means for this problem, just that it is a property that some numbers have and others do not.

Write the converse and the contrapositive of the statement, saying which is which. Note: the original statement claims that an implication is true for all *n*, and it is that implication that we are taking the converse and contrapositive of.

Write the negation of the original statement. What would you need to show to prove that the statement is false?

Even though you don't know whether 10 is solitary (in fact, nobody knows this), is the statement "if 10 is prime, then 10 is solitary" true or false?

Explain

It turns out that 8 is solitary. Does this tell you anything about the truth or falsity of the original statement, its converse or its contrapositive?

Explain

Assuming that the original statement is true, what can you say about the relationship between the *set P* of prime numbers and the *set S* of solitary numbers? Explain.

1.3 SETS

The most fundamental objects we will use in our studies (and really in all of math) are *sets*. Much of what follows might be review, but it is very important that you are fluent in the language of set theory. Most of the notation we use below is standard, although some might be a little different than what you have seen before.

For us, a set will simply be an unordered collection of objects. Two examples: we could consider the set of all actors who have played *The Doctor* on *Doctor Who,* or the set of natural numbers between 1 and 10 inclusive. In the first case, Tom Baker is an element (or member) of the set, while Idris Elba, among many others, is not an element of the set. Also, the two examples are of different sets. Two sets are equal exactly if they contain the exact same elements. For example, the set containing all of the vowels in the declaration of independence is precisely the same set as the set of vowels in the word "questionably" (namely, all of them); we do not care about order or repetitions, just whether the element is in the set or not.

Notation

We need some notation to make talking about sets easier. Consider,

$$A = \{1, 2, 3\}.$$

This is read, "A is the set containing the elements 1, 2 and 3." We use curly braces "{, }" to enclose elements of a set. Some more notation:

$$a \in \{a, b, c\}.$$

The symbol "∈ " is read "is in" or "is an element of." Thus the above means that a is an element of the set containing the letters a, b, and c. Note that this is a true statement. It would also be true to say that d is not in that set:

$$d \notin \{a, b, c\}.$$

Be warned: we write "$x \in A$" when we wish to express that one of the elements of the set A is x. For example, consider the set,

$$A - \{1, \mathrm{b}, \{x, y, z\}, \emptyset\}.$$

This is a strange set, to be sure. It contains four elements: the number 1, the letter b, the set $\{x, y, z\}$, and the empty set $\emptyset=\{\}$, the set containing no elements. Is x in A? The answer is no. None of the four elements in A are the letter x, so we must conclude that $x \in A$. Similarly, consider the set $B = \{1, \mathrm{b}\}$. Even though the elements of B are elements of A, we cannot say that the set B is one of the elements of A.

Therefore $B \notin A$. (Soon we will see that B is a subset of A, but this is different from being an *element* of A.)

We have described the sets above by listing their elements. Sometimes this is hard to do, especially when there are a lot of elements in the set (perhaps infinitely many). For instance, if we want A to be the set of all even natural numbers, would could write,

$$A = \{0, 2, 4, 6, \ldots\},$$

but this is a little imprecise. A better way would be

$$A = \{x \in \mathbb{N} : \exists n \in \mathbb{N}(x = 2n)\}.$$

Let's look at this carefully. First, there are some new symbols to digest: "$\mathbb{N}$" is the symbol usually used to denote that natural numbers, which we will take to be the set {0, 1, 2, 3, ...} Next, the colon, ":", is read *such that;* it separates the elements that are in the set from the condition that the elements in the set must satisfy. So putting this all together, we would read the set as, "the set of all x in the natural numbers, such that there exists some *n* in the natural numbers for which x is twice *n*." In other words, the set of all natural numbers, that are even. Here is another way to write the same set.

$$A = \{x \in \mathbb{N} : x \text{ is even}\}.$$

Note: Sometimes mathematicians use | or $\mathbb{N}$ for the "such that" symbol instead of the colon. Also, there is a fairly even split between mathematicians about whether 0 is an element of the natural numbers, so be careful there.

This notation is usually called set builder notation. It tells us how to build a set by telling us precisely the condition elements must meet to gain access (the condition is the logical statement after the ":" symbol). Reading and comprehending sets written in this way takes practice. Here are some more examples:

Example 1.3.1

Describe each of the following sets both in words and by listing out enough elements to see the pattern.

$$\{x : x + 3 \in \mathbb{N}\}.$$

$$\{x \in \mathbb{N} : x + 3 \in \mathbb{N}\}.$$

$$\{x : x \in \mathbb{N} \vee -x \in \mathbb{N}\}.$$

$$\{x : x \in \mathbb{N} \wedge -x \in \mathbb{N}\}.$$

Solution

This is the set of all numbers which are 3 less than a natural number (i.e., that if you add 3 to them, you get a natural number). The set could also be written as

$\{-3, -2, -1, 0, 1, 2, \ldots\}$ (note that 0 is a natural number, so -3 is in this set because $-3 + 3 = 0$).

This is the set of all natural numbers which are 3 less than a natural number. So here we just have $\{0, 1, 2, 3 \ldots\}$.

This is the set of all integers (positive and negative whole numbers, written $\mathbb{Z}$). In other words, $\{\ldots, -2, -1, 0, 1, 2, \ldots\}$.

Here we want all numbers x such that x and $-x$ are natural numbers. There is only one: 0. So we have the set $\{0\}$.

There is also a subtle variation on set builder notation. While the condition is generally given after the "such that", sometimes it is hidden in the first part. Here is an example.

Example 1.3.2

List a few elements in the sets below and describe them in words. The set $\mathbb{Z}$ is the set of integers; positive and negative whole numbers.

$$A = \{x \in \mathbb{Z} : x^2 \in \mathbb{N}\}$$

$$B = \{x^2 : x \in \mathbb{N}\}$$

Solution

The set of integers that pass the condition that their square is a natural number. Well, every integer, when you square it, gives you a non-negative integer, so a natural number. Thus $A = \mathbb{Z} = \{\ldots, -2, -1, 0, 1, 2, 3, \ldots\}$.

Here we are looking for the set of all x^2s where x is a natural number. So this set is simply the set of perfect squares. $B = \{0, 1, 4, 9, 16, \ldots\}$.

Another way we could have written this set, using more strict set builder notation, would be as $B = \{x \in \mathbb{N} : x = n^2 \text{ for some } n \in \mathbb{N}\}$.

We already have a lot of notation, and there is more yet. Below is a handy chart of symbols. Some of these will be discussed in greater detail as we move forward.

Special sets

$\emptyset$ The empty set is the set which contains no elements.

$\mathbb{U}$ The universe set is the set of all elements.

$\mathbb{N}$ The set of natural numbers. That is, $\mathbb{N} = \{0, 1, 2, 3 \ldots\}$.

$\mathbb{Z}$ The set of integers. That is, $\mathbb{Z} = \{\ldots, -2, -1, 0, 1, 2, 3, \ldots\}$.

$\mathbb{Q}$ The set of rational numbers.

$\mathbb{R}$ The set of real numbers.

$P(A)$ The power set of any set A is the set of all subsets of A.

Set Theory Notation

{, } We use these braces to enclose the elements of a set. So {1, 2, 3} is the set containing 1, 2, and 3.

: $\{x : x > 2\}$ is the set of all x such that x is greater than 2.

- $\in$: $2 \in \{1, 2, 3\}$ asserts that 2 is an element of the set {1, 2, 3}.
- $\notin$: $4 \notin < \{1, 2, 3\}$ because 4 is not an element of the set {1, 2, 3}.
- $\subseteq$: $A \subseteq B$ asserts that A is a subset of B: every element of A is also an element of B.
- $\subset$: $A \subset B$ asserts that A is a proper subset of B: every element of A is also an element of B, but $A \neq B$.
- $\cap$: $A \cap B$ is the intersection of A and B: the set containing all elements which are elements of both A and B.
- $\cup$: $A \cup B$ is the union of A and B: is the set containing all elements which are elements of A or B or both.
- $\times$: $A \times B$ is the Cartesian product of A and B: the set of all ordered pairs (a, b) with $a \in A$ and $b \in B$.
- $\setminus$: $A \setminus B$ is set difference between A and B: the set containing all elements of A which are not elements of B.
- $\overline{A}$: The complement of A is the set of everything which is not an element of A.
- $|A|$: The cardinality (or size) of A is the number of elements in A.

Investigate!

Find the cardinality of each set below.

$A = \{3, 4, \ldots, 15\}$.

$B = \{n \in \mathbb{N} : 2 < n \leq 200\}$.

$C = \{n \leq 100 : n \in \mathbb{N} \wedge \exists m \in \mathbb{N}(n = 2m + 1)\}$.

Find two sets A and B for which $|A| = 5$, $|B| = 6$, and $|A \cup B| = 9$. What is $|A \cap B|$?

Find sets A and B with $|A| = |B|$ such that $|A \cup B| = 7$ and $|A \cap B| = 3$. What is $|A|$?

Let $A = \{1, 2, \ldots, 10\}$. Define $B_2 = \{B \subseteq A : |B| = 2\}$. Find $|B_2|$.

For any sets A and B, define $AB = \{ab : a \in A \wedge b \in B\}$. If $A = \{1, 2\}$ and $B = \{2, 3, 4\}$, what is $|AB|$? What is $|A \times B|$?

Attempt the above activity before proceeding

We have already said what it means for two sets to be equal: they have exactly the same elements. Thus, for example,

$$\{1, 2, 3\} = \{2, 1, 3\}.$$

(Remember, the order the elements are written down in does not matter.) Also, since these are all ways to write the set containing the first three positive integers (how we write them doesn't matter, just what they are).

$$\{1, 2, 3\} = \{1, 1 + 1, 1 + 1 + 1\} = \{I, II, III\} = \{1, 2, 3, 1 + 2\}$$

Relationships Between Sets

What about the sets $A = \{1, 2, 3\}$ and $B = \{1, 2, 3, 4\}$? Clearly $A \neq B$, but notice that every element of A is also an element of B. Because of this we say that A is a subset of B, or in symbols $A \subset B$ or $A \subseteq B$. Both symbols are read "is a subset of." The difference is that sometimes we want to say that A is either equal to or is a subset of B, in which case we use $\subseteq$. This is analogous to the difference between $<$ and $\leq$.

Example 1.3.3

Let $A = \{1, 2, 3, 4, 5, 6\}$, $B = \{2, 4, 6\}$, $C = \{1, 2, 3\}$ and $D = \{7, 8, 9\}$.

Determine which of the following are true, false, or meaningless.

$A \subset B$.

$B \subset A$.

$B \in C$.

$\emptyset \in A$.

$\emptyset \subset A$.

$A < D$.

$3 \in C$.

$3 \subset C$.

$\{3\} \subset C$.

Solution

False. For example, $1 \in A$ but $1 \notin B$.

True. Every element in B is an element in A.

False. The elements in C are 1, 2, and 3. The *set* B is not equal to 1, 2, or 3.

False. A has exactly 6 elements, and none of them are the empty set.

True. Everything in the empty set (nothing) is also an element of A. Notice that the empty set is a subset of every set.

Meaningless. A set cannot be less than another set.

True. 3 is one of the elements of the set C.

Meaningless. 3 is not a set, so it cannot be a subset of another set.

True. 3 is the only element of the set $\{3\}$, and is an element of C, so every element in $\{3\}$ is an element of C.

In the example above, B is a subset of A. You might wonder what other sets are subsets of A. If you collect all these subsets of A into a new set, we get a set of sets. We call the set of all subsets of A the power set of A, and write it $P(A)$.

Example 1.3.4

Let $A = \{1, 2, 3\}$. Find $P(A)$.

Solution

$P(A)$ is a set of sets, all of which are subsets of A. So

$$P(A) = \{\emptyset, \{1\}, \{2\}, \{3\}, \{1, 2\}, \{1, 3\}, \{2, 3\}, \{1, 2, 3\}\}.$$

Notice that while $2 \in A$, it is wrong to write $2 \in P(A)$ since none of the elements in $P(A)$ are numbers! On the other hand, we do have $\{2\} \in P(A)$ because $\{2\} \subseteq A$.

What does a subset of $P(A)$ look like? Notice that $\{2\} \not\subseteq P(A)$ because not everything in $\{2\}$ is in $P(A)$. But we do have $\{\{2\}\} \subseteq P(A)$. The only element of $\{\{2\}\}$ is the set $\{2\}$ which is also an element of $P(A)$. We could take the collection of all subsets of $P(A)$ and call that $P(P(A))$. Or even the power set of that set of sets of sets.

Another way to compare sets is by their size. Notice that in the example above, A has 6 elements and B, C, and D all have 3 elements. The size of a set is called the set's **cardinality** . We would write $|A|= 6$, $|B|= 3$, and so on. For sets that have a finite number of elements, the cardinality of the set is simply the number of elements in the set. Note that the cardinality of $\{1, 2, 3, 2, 1\}$ is 3. We do not count repeats (in fact, $\{1, 2, 3, 2, 1\}$ is exactly the same set as $\{1, 2, 3\}$). There are sets with infinite cardinality, such as $\mathbb{N}$, the set of rational numbers (written $\mathbb{Q}$), the set of even natural numbers, and the set of real numbers ($\mathbb{R}$). It is possible to distinguish between different infinite cardinalities, but that is beyond the scope of this text. For us, a set will either be infinite, or finite; if it is finite, the we can determine its cardinality by counting elements.

Example 1.3.5

Find the cardinality of $A = \{23, 24, \ldots, 37, 38\}$.

Find the cardinality of $B = \{1, \{2, 3, 4\}, \emptyset\}$.

If $C = \{1, 2, 3\}$, what is the cardinality of $P(C)$?

Solution

Since $38 - 23 = 15$, we can conclude that the cardinality of the set is $|A| = 16$ (you need to add one since 23 is included).

Here $|B| = 3$. The three elements are the number 1, the set {2, 3, 4}, and the empty set.

We wrote out the elements of the power set $\Pi(C)$ above, and there are 8 elements (each of which set.) So $|\Pi(C)| = 8$. (You might wonder if there is a relationship between $|A|$ and $|P(A)|$ for all sets A. This is a good question which we will return to in Chapter 1.)

Operations On Sets

Example 1.3.5

Is it possible to add two sets? Not really, however there is something similar. If we want to combine two sets to get the collection of objects that are in either set, then we can take the **union** of the two sets. Symbolically,

$$C = A \cup B,$$

us, a set will either be infinite, or finite; if it is finite, the we can determine its cardinality by counting elements.

read, "C is the union of A and B," means that the elements of C are exactly the elements which are either an element of A or an element of B (or an element of both). For example, if $A = \{1, 2, 3\}$ and $B = \{2, 3, 4\}$, then $A \cup B = \{1, 2, 3, 4\}$.

The other common operation on sets is intersection. We write,

$$C = A \cap B$$

and say, "C is the intersection of A and B," when the elements in C are precisely those both in A and in B. So if $A = \{1, 2, 3\}$ and $B = \{2, 3, 4\}$, then $A \cap B = \{2, 3\}$.

Often when dealing with sets, we will have some understanding as to what "everything" is. Perhaps we are only concerned with natural numbers. In this case we would say that our universe is N. Sometimes we denote this universe by Y. Given this context, we might wish to speak of all the elements which are not in a particular set. We say B is the complement of A, and write,

$$B = \overline{A}$$

when B contains every element not contained in A. So, if our universe is $\{1, 2, \ldots, 9, 10\}$, and $A = \{2, 3, 5, 7\}$, then $\overline{A} = \{1, 4, 6, 8, 9, 10\}$.

Of course we can perform more than one operation at a time. For example, consider

$$A \cap \overline{B}.$$

This is the set of all elements which are both elements of A and not elements of B. What have we done? We've started with A and removed all of the elements which were in B. Another way to write this is the **set difference:**

$$A \cap \overline{B} = A \setminus B.$$

It is important to remember that these operations (union, intersection, complement, and difference) on sets produce other sets. Don't confuse these with the symbols from the previous section (element of and subset of). $A \cap B$ is a set, while $A \subseteq B$ is true or false. This is the same difference as between 3 + 2 (which is a number) and 3 ≤ 2 (which is false).

Example 0.3.6

Let A = {1, 2, 3, 4, 5, 6}, B = {2, 4, 6}, C = {1, 2, 3} and D = {7, 8, 9}.

If the universe is U = {1, 2, . . . , 10}, find:

$A \cup B$.

$A \cap B$.

$B \cap C$.

$A \cap D$.

$\overline{B \cup C}$.

$A \setminus B$.

$(D \cap \overline{C}) \cup \overline{A \cap B}$.

$\emptyset \cup C$.

$\emptyset \cap C$.

Solution

$A \cup B = \{1, 2, 3, 4, 5, 6\} = A$ since everything in B is already in A.

$A \cap B = \{2, 4, 6\} = B$ since everything in B is in A.

$B \cap C = \{2\}$ as the only element of both B and C is 2.

$A \cap D = \emptyset$ since A and D have no common elements.

- $\overline{B \cup C} = \{5, 7, 8, 9, 10\}$. First we find that $B \cup C$ = {1, 2, 3, 4, 6}, then we take everything not in that set.
- $A \setminus B$ = {1, 3, 5} since the elements 1, 3, and 5 are in A but not in B. This is the same as $A \cap \overline{B}$.
- $(D \cap \overline{C}) \cup \overline{A \cap B} = \{1, 3, 5, 7, 8, 9, 10\}$. The set contains all elements that are either in D but not in C (i.e., {7, 8, 9}), or not in both A and B (i.e., {1, 3, 5, 7, 8, 9, 10}).
- $\emptyset \cup C = C$ since nothing is added by the empty set.
- $\emptyset \cap C = \emptyset$ since nothing can be both in a set and in the empty set.

Having notation like this is useful. We will often want to add or remove elements from sets, and our notation allows us to do so precisely.

Example 1.3.7

If $A = \{1, 2, 3\}$, then we can describe the set we get by adding the number 4 as $A \cup \{4\}$. If we want to express the set we get by removing the number 2 from A we can do so by writing $A \setminus \{2\}$.

Careful though. If you add an element to the set, you get a new set! So you would have $B = A \cup \{4\}$ and then correctly say that B contains 4, but A does not.

You might notice that the symbols for union and intersection slightly resemble the logic symbols for "or" and "and." This is no accident. What does it mean for x to be an element of $A \cup B$? It means that x is an element of A or x is an element of B (or both). That is,

$$x \in A \cup B \quad \Leftrightarrow \quad x \in A \vee x \in B.$$

Similarly,

$$x \in A \cap B \quad \Leftrightarrow \quad x \in A \wedge x \in B$$

Also,

$$x \in \overline{A} \quad \Leftrightarrow \quad \neg(x \in A).$$

which says x is an element of the complement of A if x is not an element of A.

There is one more way to combine sets which will be useful for us: the Cartesian product, $A\ X\ B$. This sounds fancy but is nothing you haven't seen before. When you graph a function in calculus, you graph it in the Cartesian plane. This is the set of all ordered pairs of real numbers (x, y) . We can do this for any pair of sets, not just the real numbers with themselves.

Put another way, $A \times B = \{(a, b) : a\ A\ b \in B\}$. The first coordinate comes from the first set and the second coordinate comes from the second set. Sometimes we will want to take the Cartesian product of a set with itself, and this is fine: A x A $= \{(a, b) : a, b \in A\}$ (we might also write A^2 for this set). Notice that in $A \times A$, we still want *all* ordered pairs, not just the ones where the first and second coordinate are the same. We can also take products of 3 or more sets, getting ordered triples, or quadruples, and so on.

Example 1.3.8

Let $A = \{1, 2\}$ and $B = \{3, 4, 5\}$. Find $A \times B$ and $A \times A$. How many elements do you expect to be in $B \times B$?

Solution

$A \times B = \{(1, 3), (1, 4), (1, 5), (2, 3), (2, 4), (2, 5)\}$.

$A \times A = A^2 = \{(1, 1), (1, 2), (2, 1), (2, 2)\}$.

$|B \times B| = 9$. There will be 3 pairs with first coordinate 3, three more with first coordinate 4, and a final three with first coordinate 5.

Venn Diagrams

There is a very nice visual tool we can use to represent operations on sets. A Venn diagram displays sets as intersecting circles. We can shade the region we are talking about when we carry out an operation. We can also represent cardinality of a particular set by putting the number in the corresponding region.

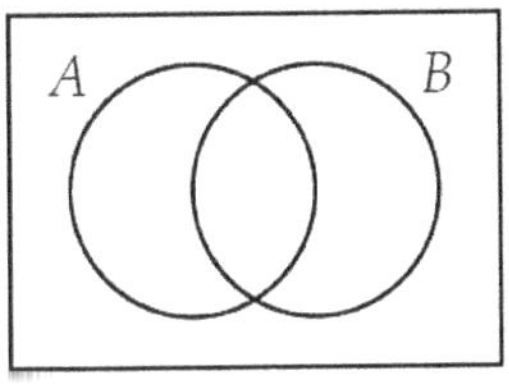

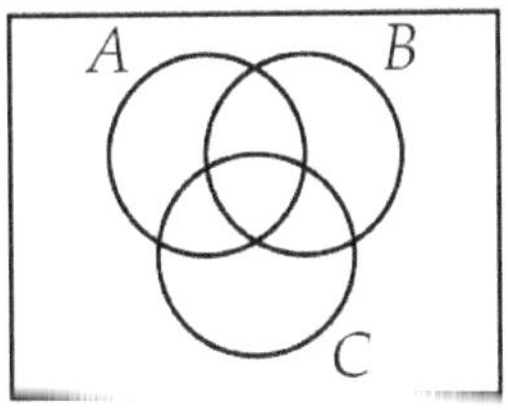

Each circle represents a set. The rectangle containing the circles represents the universe. To represent combinations of these sets, we shade the corresponding region. For example, we could draw $A \cap B$ as:

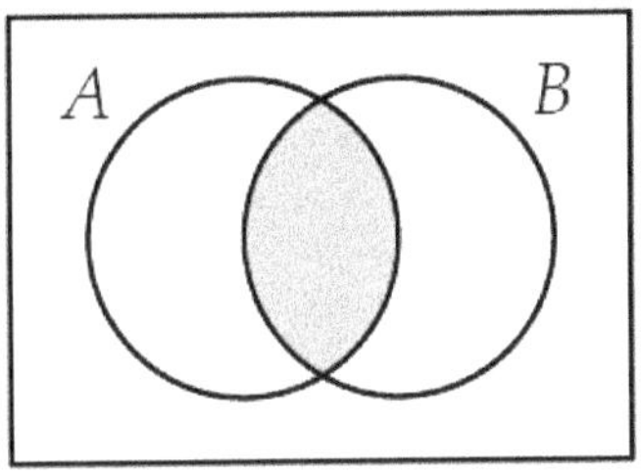

Here is a representation of $A \cap \overline{B}$, or equivalently $A \setminus B$:

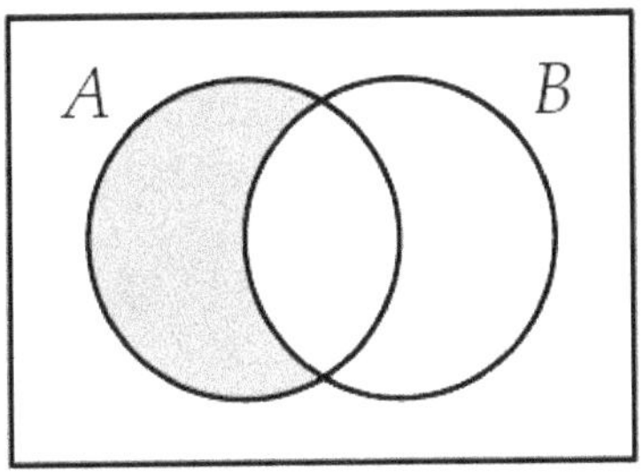

A more complicated example is $(B \cap C) \cup (C \cap \overline{A})$, as seen below.

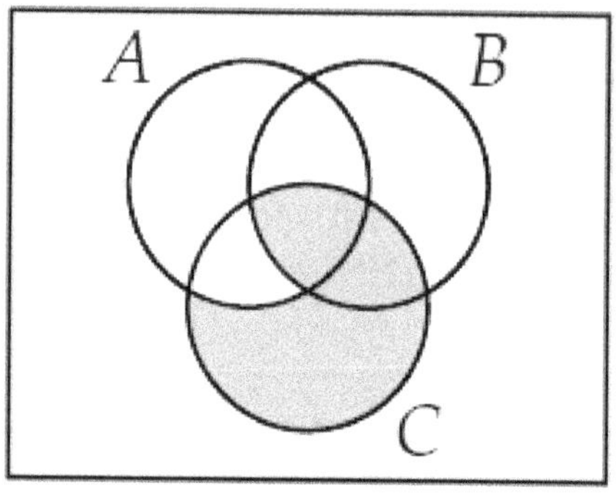

Notice that the shaded regions above could also be arrived at in another way. We could have started with all of C, then excluded the region where C and A overlap outside of B. That region is $(A \cap C) \cap \overline{B}$. So the above Venn diagram also represents $C \cap \overline{\left((A \cap C) \cap \overline{B}\right)}$. So using just the picture,

we have determined that

$$(B \cap C) \cup (C \cap \overline{A}) = C \cap \overline{\left((A \cap C) \cap \overline{B}\right)}.$$

Exercises

Let $A = \{1, 4, 9\}$ and $B = \{1, 3, 6, 10\}$. Find each of the following sets.

$A \cup B$.

$A \cap B$.

$A \setminus B$.

$B \setminus A$.

Find the least element of each of the following sets, if there is one.

$\{n \in \mathbb{N} : n^2 - 3 \geq 2\}$.

$\{n \in \mathbb{N} : n^2 - 5 \in \mathbb{N}\}$.

$\{n^2 + 1 : n \in \mathbb{N}\}$.

$\{n \in \mathbb{N} : n = k^2 + 1 \text{ for some } k \in \mathbb{N}\}$.

Find the following cardinalities:

$|A|$ when $A = \{4, 5, 6, \ldots, 37\}$.

$|A|$ when $A = \{x \in \mathbb{Z} : -2 \leq x \leq 100\}$.

$|A \cap B|$ when $A = \{x \in \mathbb{N} : x \leq 20\}$ and $B = \{x \in \mathbb{N} : x \text{ is prime}\}$.

Find a set of largest possible size that is a subset of both $\{1, 2, 3, 4, 5\}$ and $\{2, 4, 6, 8, 10\}$.

Find a set of smallest possible size that has both {1, 2, 3, 4, 5} and {2, 4, 6, 8, 10} as subsets.

Let A = $\{n \in \mathbb{N} : 20\ n \leq 50\}$ and $B = \{n \in \mathbb{N} : 10 < n \leq 30\}$. Suppose C is a set such that $C \not\subset A$ and $C \not\subset B$. What is the largest possible cardinality of C?

Let $A = \{1, 2, 3, 4, 5\}$ and $B = \{2, 3, 4\}$. How many sets C have the property that $C \subseteq A$ and $B \subseteq C$.

et $A = \{1, 2, 3, 4, 5\}$, $B = \{3, 4, 5, 6, 7\}$, and $C = \{2, 3, 5\}$.

Find $A \cap B$.

Find $A \cup B$.

Find $A \setminus B$.

Find $A \cap \overline{(B \cup C)}$.

Let $A = \{x \in \mathbb{N} : 4 \leq x < 12\}$ and $B = \{x \in \mathbb{N} : x \text{ is even}\}$.

Find $A \cap B$.

Find $A \setminus B$.

Let $A = \{x \in \mathbb{N} : 3 \leq x \leq 13\}$, $B = \{x \in \mathbb{N} : x \text{ is even}\}$, and $C = \{x \in \mathbb{N} : x \text{ is odd}\}$.

Find $A \cap B$.

Find $A \cup B$.

Find $B \cap C$.

Find $B \cup C$.

Find an example of sets A and B such that $A \cap B = \{3, 5\}$ and $A \cup B = \{2, 3, 5, 7, 8\}$.

Find an example of sets A and B such that $A \subseteq B$ and $A \in B$.

Recall $\mathbb{Z} = \{\ldots, 2, 1, 0, 1, 2, \ldots\}$ (the integers). Let $\mathbb{Z}^+ = \{1, 2, 3, \ldots\}$ be the positive integers. Let $2\mathbb{Z}$ be the even integers, $3\mathbb{Z}$ be the multiples of 3, and so on.

Is $\mathbb{Z}^+ \subseteq 2\mathbb{Z}$? Explain.

Is $2\mathbb{Z} \subseteq \mathbb{Z}^+$? Explain.

Find $2\mathbb{Z} \cap 3\mathbb{Z}$. Describe the set in words, and using set notation.

Express $\{x \in \mathbb{Z} : \exists y \in Z\ (x = 2y \vee x = 3y)\}$ as a union or intersection of two sets already described in this problem.

Let A_2 be the set of all multiples of 2 except for 2. Let A_3 be the set of all multiples of 3 except for 3. And so on, so that A_n is the set of all multiples of n except for n, for any $n \geq 2$. Describe (in words) the set $\overline{A_2 \cup A_3 \cup A_4 \cup \cdots}$.

Draw a Venn diagram to represent each of the following:

$A \cup \overline{B}$

$\overline{(A \cup B)}$

$A \cap (B \cup C)$

$(A \cap B) \cup C$

$\overline{A} \cap B \cap \overline{C}$

$(A \cup B) \setminus C$

Describe a set in terms of A and B (using set notation) which has the following Venn diagram:

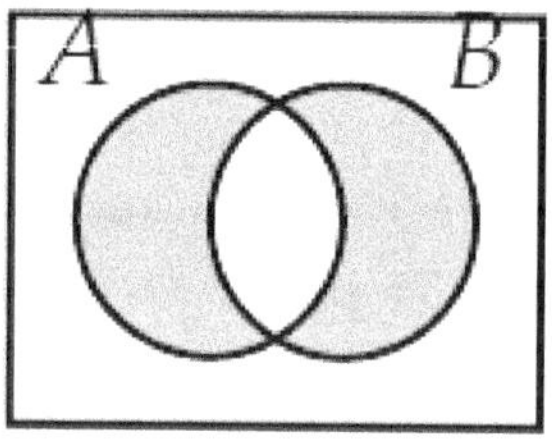

Let $A = \{a, b, c, d\}$. Find $\Pi(A)$.

Let $A = \{1, 2, \ldots, 10\}$. How many subsets of A contain exactly one element (i.e., how many singleton subsets are there)? singleton set

How many doubleton subsets (containing exactly two elements) are there? doubleton set

Let $A = \{1, 2, 3, 4, 5, 6\}$. Find all sets $B \in \Pi(A)$ which have the property $\{2, 3, 5\} \subseteq B$.

Find an example of sets A and B such that $|A| = 4$, $|B| = 5$, and $|A \cup B| = 9$.

Find an example of sets A and B such that $|A| = 3$, $|B| = 4$, and $|A \cup B| = 5$.

Are there sets A and B such that $|A| = |B|$, $|A \cup B| = 10$, and $|A \cap B| = 5$? Explain.

Let $A = \{2, 4, 6, 8\}$. Suppose B is a set with $|B| = 5$.

What are the smallest and largest possible values of $|A \cup B|$? Explain.

What are the smallest and largest possible values of $|A \cap B|$? Explain.

What are the smallest and largest possible values of $|A \times B|$? Explain.

Let $X = \{n \in \mathbb{N} : 10 < n\ 20\}$. Find examples of sets with the properties below and very briefly explain why your examples work.

A set $A \subseteq \mathbb{N}$ with $|A| = 10$ such that $X \setminus A = \{10, 12, 14\}$.

A set $B \in \Pi(X)$ with $|B| = 5$.

A set $C \subseteq \Pi(X)$ with $|C| = 5$.

A set $D \subseteq X \times X$ with $|D| = 5$

A set $E \subseteq X$ such that $|E| \in E$.

Let A, B and C be sets.

Suppose that $A \subseteq B$ and $B \subseteq C$. Does this mean that $A \subseteq C$? Prove your answer. Hint: to prove that $A \subseteq C$ you must prove the implication, "for all x, if $x \in A$ then $x \in C$."

Suppose that $A \in B$ and $B \in C$. Does this mean that $A \in C$? Give an example to prove that this does NOT always happen (and explain why your example works). You should be able to give an example where $|A| = |B| = |C| = 2$.

In a regular deck of playing cards there are 26 red cards and 12 face cards. Explain, using sets and what you have learned about cardinalities, why there are only 32 cards which are either red or a face card.

Find an example of a set A with $|A| = 3$ which contains only other sets and has the following property: for all sets $B \in A$, we also have $B \in A$. Explain why your example works. (FYI: sets that have this property are called **transitive.**)

Consider the sets A and B, where $A = \{3, |B|\}$ and $B = \{1, |A|, |B|\}$. What are the sets?

Explain why there is no set A which satisfies $A = \{2, |A|\}$.

Find all sets A, B, and C which satisfy the following.

$$A = \{1, |B|, |C|\}$$
$$B = \{2, |A|, |C|\}$$
$$C = \{1, 2, |A|, |B|\}.$$

1.4 FUNCTIONS

A **function** is a rule that assigns each input exactly one output. We call the output the image of the input. The set of all inputs for a function is called the domain. The set of all allowable outputs is called the codomain. We would write $f : X \to Y$ to describe a function with name f, domain X and codomain Y. This does not tell us which *function* f is though. To define the function, we must describe the rule. This is often done by giving a formula to compute the output for any input (although this is certainly not the only way to describe the rule).

For example, consider the function $f : \mathbb{N} \to \mathbb{N}$ defined by $f(x) = x^2 + 3$. Here the domain and codomain are the same set (the natural numbers).

The rule is: take your input, multiply it by itself and add 3. This works because we can apply this rule to every natural number (every element of the domain) and the result is always a natural number (an element of the codomain). Notice though that not every natural number is actually an output (there is no way to get 0, 1, 2, 5, etc.). The set of natural numbers that are outputs is called the range of the function (in this case, the range is $\{3, 4, 7, 12, 19, 28, \ldots,\}$ all the natural numbers that are 3 more than a perfect square).

The key thing that makes a rule a function is that there is exactly one output for each input. That is, it is important that the rule be a good rule. What output do we assign to the input 7? There can only be one answer for any particular function.

Example 1.4.1

The following are all examples of functions:

$f : \mathbb{Z} \to \mathbb{Z}$ defined by $f(n) = 3n$. The domain and codomain are both the set of integers. However, the range is only the set of integer multiples of 3.

g: $\{1, 2, 3\} \to \{a, b, c\}$ defined by $g(1) = c$, $g(2) = a$ and $g(3) = a$. The domain is the set $\{1, 2, 3\}$, the codomain is the set $\{a, b, c\}$ and the range is the set $\{a, c\}$. Note that $g(2)$ and $g(3)$ are the same element of the codomain. This is okay since each element in the domain still has only one output.

$h : \{1, 2, 3, 4\} \to$ N defined by the table:

x	1	2	3	4
$h(x)$	3	6	9	12

Here the domain is the finite set 1, 2, 3, 4 and to codomain is the set of natural numbers, N. At first you might think this function is the same as f defined above. It is absolutely not. Even though the rule is the same, the domain and codomain are different, so these are two different functions.

Example 1.4.2

Just because you can describe a rule in the same way you would write a function, does not mean that the rule is a function. The following are NOT functions.

$f : \mathbb{N} \to \mathbb{N}$ defined by $f(n) = \frac{n}{2}$. The reason this is not a function is because not every input has an output. Where does f send 3? The rule says that $f(3) = \frac{3}{2}$, but $\frac{3}{2}$ is not an element of the codomain.

Consider the rule that matches each person to their phone number. If you think of the set of people as the domain and the set of phone numbers as the codomain, then this is not a function, since some people have two phone numbers. Switching the domain and codomain sets doesn't help either, since some phone numbers belong to multiple people (assuming some households still have landlines when you are reading this).

Describing Functions

It is worth making a distinction between a function and its description. The function is the abstract mathematical object that in some way exists whether or not anyone ever talks about it. But when we do want to talk about the function, we need a way to describe it. A particular function can be described in multiple ways.

Some calculus textbooks talk about the *Rule of Four,* that every function can be described in four ways: algebraically (a formula), numerically (a table), graphically, or in words. In discrete math, we can still use any of these to describe functions, but we can also be more specific since we are primarily concerned with functions that have N or a finite subset of N as their domain.

Describing a function graphically usually means drawing the graph of the function: plotting the points on the plane. We can do this, and might get a graph like the following for a function $f: \{1, 2, 3\} \to \{1, 2, 3\}$.

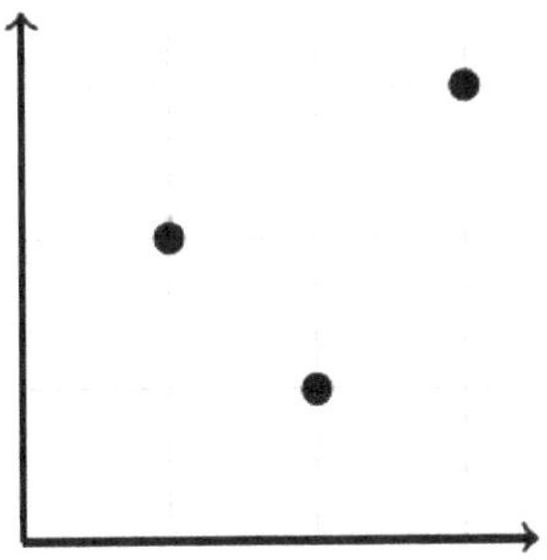

It would be absolutely WRONG to connect the dots or try to fit them to some curve. There are only three elements in the domain. A curve would mean that the domain contains an entire interval of real numbers.

Here is another way to represent that same function:

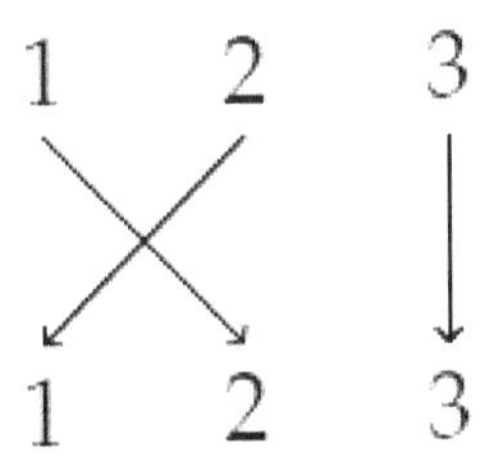

This shows that the function f sends 1 to 2, 2 to 1 and 3 to 3: just follow the arrows.

The arrow diagram used to define the function above can be very helpful in visualizing functions. We will often be working with functions with finite domains, so this kind of picture is often more useful than a traditional graph of a function.

Note that for finite domains, finding an algebraic formula that gives the output for any input is often impossible. Of course we could use a piecewise defined function, like

$$f(x) = \begin{cases} x+1 & \text{if } x = 1 \\ x-1 & \text{if } x = 2 \\ x & \text{if } x = 3 \end{cases}.$$

This describes exactly the same function as above, but we can all agree is a ridiculous way of doing so.

Since we will so often use functions with small domains and codomains, let's adopt some notation to describe them. All we need is some clear way of denoting the image of each element in the domain. In fact, writing a table of values would work perfectly:

x	0	1	2	3	4
$f(x)$	3	3	2	4	1

We simplify this further by writing this as a "matrix" with each input directly over its output:

$$f = \begin{pmatrix} 0 & 1 & 2 & 3 & 4 \\ 3 & 3 & 2 & 4 & 1 \end{pmatrix}.$$

Note this is just notation and not the same sort of matrix you would find in a linear algebra class (it does not make sense to do operations with these matrices, or row reduce them, for example).

One advantage of the two-line notation over the arrow diagrams is that it is harder to accidentally define a rule that is not a function using two-line notation.

Example 1.4.3

Which of the following diagrams represent a function? Let $X = \{1, 2, 3, 4\}$ and $Y = \{a, b, c, d\}$.

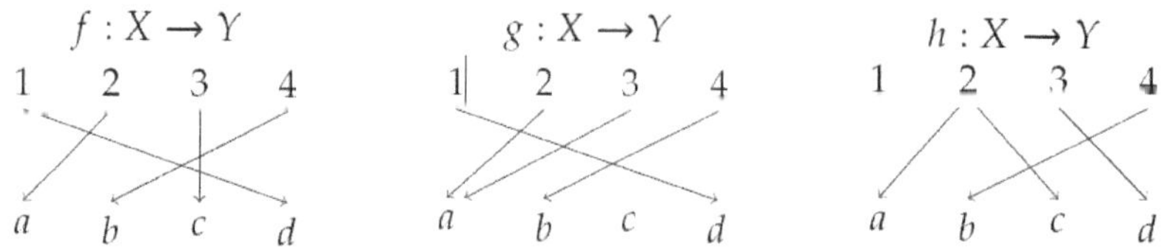

Solution

f is a function. So is g. There is no problem with an element of the codomain not being the image of any input, and there is no problem with a from the codomain being the image of both 2 and 3 from the domain. We could use our two-line notation to write these as

$$f = \begin{pmatrix} 1 & 2 & 3 & 4 \\ d & a & c & b \end{pmatrix} \qquad g = \begin{pmatrix} 1 & 2 & 3 & 4 \\ d & a & a & b \end{pmatrix}.$$

However, *h* is NOT a function. In fact, it fails for two reasons. First, the element 1 from the domain has not been mapped to any element from the codomain. Second, the element 2 from the domain has been mapped to more than one element from the codomain (a and c). Note that either one of these problems is enough to make a rule not a function. In general, neither of the following mappings are functions:

It might also be helpful to think about how you would write the two-line notation for *h*. We would have something like:

$$h = \begin{pmatrix} 1 & 2 & 3 & 4 \\ & a, c? & d & b \end{pmatrix}.$$

There is nothing under 1 (bad) and we needed to put more than one thing under 2 (very bad). With a rule that is actually a function, the two-line notation will always "work".

We will also be interested in functions with domain ℕ. Here two-line notation is no good, but describing the function algebraically is often possible. Even tables are a little awkward, since they do not describe the function completely. For example, consider the function $f\colon \mathbb{N} \to \mathbb{N}$ given by the table below.

x	0	1	2	3	4	5	...
$f(x)$	0	1	4	9	16	25	...

Have I given you enough entries for you to be able to determine $f(6)$? You might guess that $f(6) = 36$, but there is no way for you to know this for sure. Maybe I am being a jerk and intended $f(6) = 42$. In fact, for every natural number n, there is a function that agrees with the table above, but for which $f(6)=n$.

Okay, suppose I really did mean for $f(6)=36$, and in fact, for the rule that you think is governing the function to actually be the rule. Then I should say what that rule is. $f(n) = n^2$. Now there is no confusion possible.

Giving an explicit formula that calculates the image of any element in the domain is a great way to describe a function. We will say that these explicit rules are **closed formulas** for the function.

There is another very useful way to describe functions whose domain is N, that rely specifically on the structure of the natural numbers. We can define a function *recursively*!

Example 1.4.4

Consider the function $f: \mathbb{N} \to \mathbb{N}$ given by $f(0) = 0$ and $f(n+1) = f(n) + 2n + 1$. Find $f(6)$.

Solution

The rule says that $f(6) = f(5) + 11$ (we are using $6 = n+1$ so $n=5$). We don't know what $f(5)$ is though. Well, we know that $f(5) = f(4) + 9$. So we need to compute $f(4)$, which will require knowing $f(3)$, which will require $f(2)$,... will it ever end?

Yes! In fact, this process will always end because we have N as our domain, so there is a least element. And we gave the value of $f(0)$ explicitly, so we are good. In fact, we might decide to work up to $f(6)$ instead of working down from $f(6)$:

$f(1) = f(0) + 1 = 0 + 1 = 1$

$f(2) = f(1) + 3 = 1 + 3 = 4$

$f(3) = f(2) + 5 = 4 + 5 = 9$

$f(4) = f(3) + 7 = 9 + 7 = 16$

$f(5) = f(4) + 9 = 16 + 9 = 25$

$f(6) = f(5) + 11 = 25 + 11 = 36$

It looks that this recursively defined function is the same as the explicitly defined function $f(n) = n^2$. Is it? Later we will prove that it is.

Recursively defined functions are often easier to create from a "real world" problem, because they describe how the values of the functions are changing. However, this comes with a price. It is harder to calculate the image of a single input, since you need to know the images of other (previous) elements in the domain.

Recursively Defined Functions

For a function $f: \mathbb{N} \to \mathbb{N}$, a recursive definition consists of an **initial condition** together with a **recurrence relation.** The initial For a function $f: \mathrm{N} \to \mathrm{N}$, a recursive definition consists of an condition is the explicitly given value of $f(0)$. The recurrence relation is a formula for $f(n + 1)$ in terms for $f(n)$ (and possibly n itself).

Example 1.4.5

Give recursive definitions for the functions described below.

$f: \mathbb{N} \rightarrow \mathbb{N}$ gives the number of snails in your terrarium n years after you built it, assuming you started with 3 snails and the number of snails doubles each year.

$g: \mathbb{N} \rightarrow \mathbb{N}$ gives the number of push-ups you do n days after you started your push-ups challenge, assuming you could do 7 push-ups on day 0 and you can do 2 more push-ups each day.

$h: \mathbb{N} \rightarrow \mathbb{N}$ defined by $h(n) = n!$. Recall that $n! = 1.2.3. \ldots\ldots (n-1)\, n$ is the product of all numbers from 1 through n. We also define $0! = 1$.

Solution

1. The initial condition is $f(0) = 3$. To get $f(n + 1)$ we would double the number of snails in the terrarium the previous year, which is given by $f(n)$. Thus $f(n + 1) = 2f(n)$. The full recursive definition contains both of these, and would be written,

$$f(0) = 3; f(n + 1) = 2f(n).$$

2. We are told that on day 0 you can do 7 push-ups, so $g(0) = 7$. The number of push-ups you can do on day $n + 1$ is 2 more than the number you can do on day n, which is given by $g(n)$.

Thus

$$g(0) = 7; g(n + 1) = g(n) + 2.$$

3.Here $h(0) = 1$. To get the recurrence relation, think about how you can get $h(n + 1) = (n + 1)!$ from $h(n) = n!$. If you write out both of these as products, you see that $(n + 1)!$ is just like n! except you have one more term in the product, an extra $n + 1$. So we have,

$$h(0) = 1; h(n + 1) = (n + 1) \cdot h(n).$$

Surjections, Injections, and Bijections

We now turn to investigating special properties functions might or might not possess.

In the examples above, you may have noticed that sometimes there are elements of the codomain which are not in the range. When this sort of the thing does not happen, (that is, when everything in the codomain is in the range) we say the function is **onto** or that the function maps the domain onto the codomain. This terminology should make sense: the function puts the domain (entirely) on top of the codomain. The fancy math term for an onto function is a **surjection,** and we say that an onto function is a **surjective** function.

In pictures:

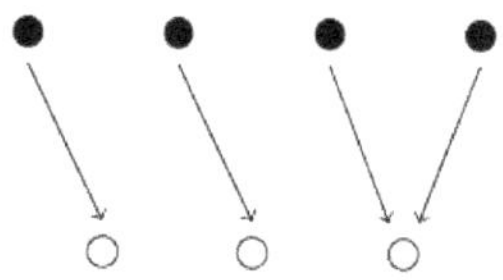

Surjective

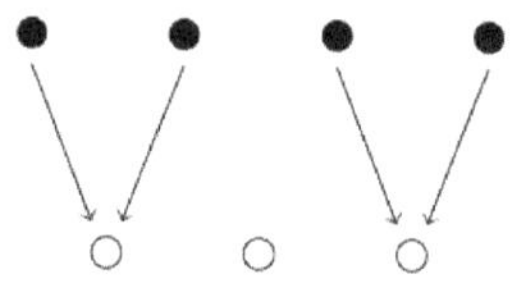

Not surjective

Example 1.4.6

Which functions are surjective (i.e., onto)?

$f: \mathbb{Z} \to \mathbb{Z}$ defined by $f(n) = 3n$.

g: {1, 2, 3) →{ a, b, c} defined by

$$g = \begin{pmatrix} 1 & 2 & 3 \\ c & a & a \end{pmatrix}.$$

h: {1, 2, 3} → {1, 2, 3} defined as follows:

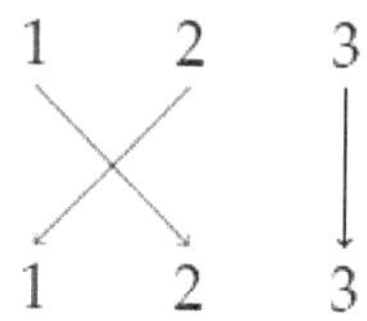

Solution

f is not surjective. There are elements in the codomain which are not in the range. For example, no $n \in \mathbb{Z}$ gets mapped to the number 1 (the rule would say that $\frac{1}{3}$ would be sent to 1, but $\frac{1}{3}$ is not in the domain). In fact, the range of the function is $3\mathbb{Z}$ (the integer multiples of 3), which is not equal to $\mathbb{Z}$.

g is not surjective. There is no $x \in \{1, 2, 3\}$ (the domain) for which g(x) = b, so b, which is in the codomain, is not in the range. Notice that there is an element from the codomain "missing" from the bottom row of the matrix.

h is surjective. Every element of the codomain is also in the range. Nothing in the codomain is missed.

To be a function, a rule cannot assign a single element of the domain to two or more different elements of the codomain. However, we have seen that the reverse is permissible: a function might assign the same element of the codomain to two or more different elements of the domain. When this does not occur (that is, when each element of the codomain is the image of at most one element of the domain) then we say the function is one-to-one. Again, this terminology makes sense:

we are sending at most one element from the domain to one element from the codomain. One input to one output. The fancy math term for a one-to-one function is an injection. We call one-to-one functions **injective** functions.

In pictures:

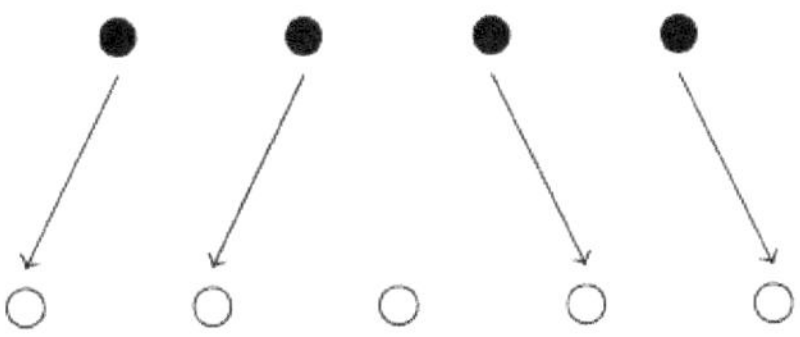

Injective

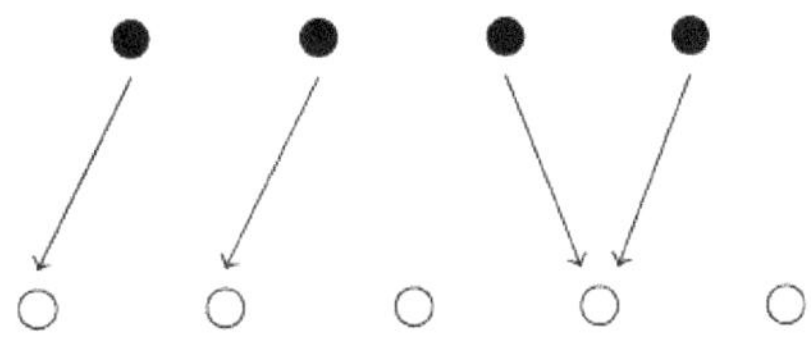

Not injective

Example 1.4.7

Which functions are injective (i.e., one-to-one)?

$f : \mathbb{Z} \to \mathbb{Z}$ defined by $f(n) = 3n$.

$g : \{1, 2, 3\} \to \{a, b, c\}$ defined by

$$g = \begin{pmatrix} 1 & 2 & 3 \\ c & a & a \end{pmatrix}.$$

$h : \{1, 2, 3\} \to \{1, 2, 3\}$ defined as follows:

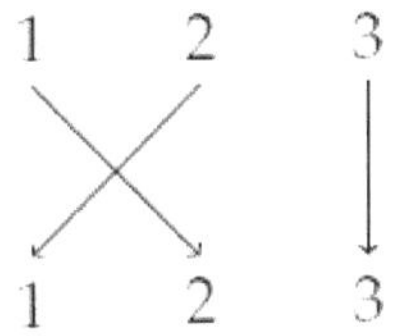

Solution

f is injective. Each element in the codomain is assigned to at most one element from the domain. If x is a multiple of three, then only $x/3$ is mapped to x. If x is not a multiple of 3, then there is no input corresponding to the output x.

g is not injective. Both inputs 2 and 3 are assigned the output a. Notice that there is an element from the codomain that appears more than once on the bottom row of the matrix.

h is injective. Each output is only an output once.

Be careful: "surjective" and "injective" are NOT opposites. You can see in the two examples above that there are functions which are surjective but not injective, injective but not surjective, both, or neither. In the case when a function is both one-to-one and onto (an injection and surjection), we say the function is a bijection, or that the function is a **bijective** function.

Injective vs Surjective

A function is **injective** provided every element of the codomain is the image of *at most* one element from the domain.

A function is **surjective** provided every element of the codomain is the image of *at least* one element from the domain.

Notice both properties are determined by what happens to elements of the codomain: they could be repeated as images or they could be "missed" (not be images). Injective functions do not have repeats but might or might not miss elements. Surjective functions do not miss elements, but might or might not have repeats. The bijective functions are those that do not have repeats and do not miss elements.

Image and Inverse Image

When discussing functions, we have notation for talking about an element of the domain (say x) and its corresponding element in the codomain (we write $f(x)$, which is the image of x). Sometimes we will want to talk about all the elements that are images of some subset of the domain. It would also be nice to start with some element of the codomain (say y) and talk about which element or elements (if any) from the domain it is the image of. We could write "those x in the domain such that $f(x) = y$," but this is a lot of writing. Here is some notation to make our lives easier.

To address the first situation, what we are after is a way to describe the set of images of elements in some subset of the domain. Suppose $f: X \to Y$ is a function and that $A \subseteq X$ is some subset of the domain (possibly all of it). We will use the notation $f(A)$ to denote the image of A under f, namely the set of elements in Y that are the image of elements from A. That is, $f(A) = \{f(a) \in Y : a \in A\}$.

We can do this in the other direction as well. We might ask which elements of the domain get mapped to a particular set in the codomain. Let $f: X \to Y$ be a function and suppose $B \subseteq Y$ is a subset of the codomain. Then we will write $f^{-1}(B)$ for the **inverse image** of B **under** f, namely the set of elements in X whose image are elements in B. In other words, $f^{-1}(B) = \{x \in X : f(x) \in B\}$.

Often we are interested in the element(s) whose image is a particular element y of in the codomain. The notation above works: $f^{-1}(\{y\})$ is the set of all elements in the domain that f sends to y. It makes sense to think of this as a set: there might not be anything sent to y (if y is not in the range), in which case $f^{-1}(\{y\}) = \emptyset$. Or f might send multiple elements to y (if f is not injective). As a notational convenience, we usually drop the set braces around the y and write $f^{-1}(y)$ instead for this set.

WARNING: $f^{-1}(y)$ is not an inverse function! Inverse functions only exist for bijections, but f^{-1} y is defined for any function f. The point: f^{-1} y is a set, not an element of the domain. This is just sloppy notation for f^{-1} y . To help make this distinction, we would call f^{-1} y the **complete inverse image of** y under f. It is not the image of y under f^{-1} (since the function f^{-1} might not exist).

Example 1.4.8

Consider the function f: {1, 2, 3, 4, 5, 6} → {a, b, c, d} given by

$$f = \begin{pmatrix} 1 & 2 & 3 & 4 & 5 & 6 \\ a & a & b & b & b & c \end{pmatrix}.$$

Find $f(\{1, 2, 3\})$, $f^{-1}(\{a, b\})$, and $f^{-1}(d)$.

Solution

$f(\{1, 2, 3\}) = \{a, b\}$ since a and b are the elements in the codomain to which f sends 1 and 2.

$f^{-1}(\{a, b\}) = \{1, 2, 3, 4, 5\}$ since these are exactly the elements that f sends to a and b.

$f^{-1}(d) = \emptyset$ since d is not in the range of f.

Example 1.4.9

Consider the function $g: \mathbb{Z} \to \mathbb{Z}$ defined by $g(n) = n^2 + 1$. Find $g(1)$ and $g(\{1\})$. Then find $g^{-1}(1)$, $g^{-1}(2)$, and $g^{-1}(3)$.

Solution

Note that $g(1) \neq g(\{1\})$. The first is an element: $g(1) = 2$. The second is a set: $g(\{1\}) = \{2\}$.

To find $g^{-1}(1)$, we need to find all integers n such that $n^2 + 1 = 1$. Clearly only 0 works, so $g^{-1}(1) = \{0\}$ (note that even though there is only one element, we still write it as a set with one element in it).

To find $g^{-1}(2)$, we need to find all n such that $n^2 + 1 = 2$. We see $g^{-1}(2) = \{-1, 1\}$.

Finally, if $n^2 + 1 = 3$, then we are looking for an n such that $n^2 = 2$. There are no such integers so $g^{-1}(3) = \emptyset$.

Since $f^{-1}(y)$ is a set, it makes sense to ask for $|f^{-1}(y)|$, the number of elements in the domain which map to y.

Example 1.4.10

Find a function f: {1, 2, 3, 4, 5} → $\mathbb{N}$ such that $|f^{-1}(7)|=5$

Solution

There is only one such function. We need five elements of the domain to map to the number $7 \in \mathbb{N}$. Since there are only five elements in the domain, all of them must map to 7. So

$$f = \begin{pmatrix} 1 & 2 & 3 & 4 & 5 \\ 7 & 7 & 7 & 7 & 7 \end{pmatrix}.$$

Function Definitions

Here is a summary of all the main concepts and definitions we use when working with functions.

A **function** is a rule that assigns each element of a set, called the **domain,** to exactly one element of a second set, called the **codomain.**

Notation: $f: X \to Y$ is our way of saying that the function is called f, the domain is the set X, and the codomain is the set Y.

To specify the rule for a function with small domain, use **two-line notation** by writing a matrix with each output directly below its corresponding input, as in:

$$f = \begin{pmatrix} 1 & 2 & 3 & 4 \\ 2 & 1 & 3 & 1 \end{pmatrix}.$$

- $f(x) = y$ means the element x of the domain (input) is assigned to the element y of the codomain. We say y is an output. Alternatively, we call y the **image** of x **under** f.
- The **range** is a subset of the codomain. It is the set of all elements which are assigned to at least one element of the domain by the function. That is, the range is the set of all outputs.
- A function is **injective** (an **injection** or **one-to-one**) if every element of the codomain is the image of **at most** one element from the domain.
- A function is surjective (a **surjection** or **onto**) if every element of the codomain is the image of **at least** one element from the domain.
- A **bijection** is a function which is both an injection and surjection. In other words, if every element of the codomain is the image of **exactly one** element from the domain.
- The **image** of an element x in the domain is the element y in the codomain that x is mapped to. That is, the image of x under f is f(x).

- The **complete inverse image** of an element y in the codomain, written $f^{-1}(y)$, is the set of all elements in the domain which are assigned to y by the function.
- The **image** of a subset A of the domain is the set $f(A) = \{f(a) \in Y : a \in A\}$.
- The **inverse image** of a subset B of the codomain is the set $f^{-1}(B) = \{x \in X : f(x) \in B\}$.

Exercises

1. **Consider the function $f : \{1, 2, 3, 4\} \to \{1, 2, 3, 4\}$ given by:**

$$f(n) = \begin{pmatrix} 1 & 2 & 3 & 4 \\ 4 & 1 & 3 & 4 \end{pmatrix}.$$

(a) Find $f(1)$.

(b) Find an element n in the domain such that $f(n) = 1$.

(c) Find an element n of the domain such that $f(n) = n$.

(d) Find an element of the codomain that is not in the range.

2. **The following functions all have {1, 2, 3, 4, 5} as both their domain and codomain. For each, determine whether it is (only) injective, (only) surjective, bijective, or neither injective nor surjective.**

(a) $f = \begin{pmatrix} 1 & 2 & 3 & 4 & 5 \\ 3 & 3 & 3 & 3 & 3 \end{pmatrix}$.

(b) $f = \begin{pmatrix} 1 & 2 & 3 & 4 & 5 \\ 2 & 3 & 1 & 5 & 4 \end{pmatrix}$.

(c) $f(x) = 6 - x$.

(d) $f(x) = \begin{cases} x/2 & \text{if } x \text{ is even} \\ (x+1)/2 & \text{if } x \text{ is odd} \end{cases}$.

3. **The following functions all have domain {1, 2, 3, 4, 5} and codomain {1, 2, 3}. For each, determine whether it is (only) injective, (only) surjective, bijective, or neither injective nor surjective.**

(a) $f = \begin{pmatrix} 1 & 2 & 3 & 4 & 5 \\ 1 & 2 & 1 & 2 & 1 \end{pmatrix}$.

(b) $f = \begin{pmatrix} 1 & 2 & 3 & 4 & 5 \\ 1 & 2 & 3 & 1 & 2 \end{pmatrix}$.

(c) $f(x) = \begin{cases} x & \text{if } x \leq 3 \\ x - 3 & \text{if } x > 3 \end{cases}$.

4. **The following functions all have domain {1, 2, 3, 4} and codomain {1, 2, 3, 4, 5}. For each, determine whether it is (only) injective, (only) surjective, bijective, or neither injective nor surjective.**

(a) $f = \begin{pmatrix} 1 & 2 & 3 & 4 \\ 1 & 2 & 5 & 4 \end{pmatrix}$. (b) $f = \begin{pmatrix} 1 & 2 & 3 & 4 \\ 1 & 2 & 3 & 2 \end{pmatrix}$.

(c) $f(x)$ gives the number of letters in the English word for the number x. For example, $f(1) = 3$ since "one" contains three letters.

5. **Write out all functions f: {1, 2, 3}→{a, b} (using two-line notation).**

   (a) How many functions are there? (b) How many are injective?

   (c) How many are surjective? (d) How many are bijective?

6. **Write out all functions f:{1, 2}→{a, b, c} (in two-line notation).**

   (a) How many functions are there? (b) How many are injective?

   (c) How many are surjective? (d) How many are bijective?

7. **Consider the function f: {1, 2, 3, 4, 5}→{ 1, 2, 3, 4} given by the table below:**

x	1	2	3	4	5
$f(x)$	3	2	4	1	2

   (a) Is f injective? Explain. (b) Is f surjective? Explain.

   (c) Write the function using two-line notation.

8. **Consider the function f: {1, 2, 3, 4}→{ 1, 2, 3, 4} given by the graph below.**

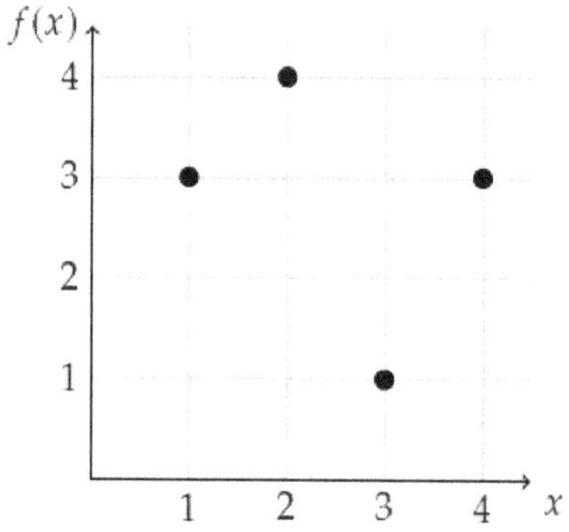

   (a) Is f injective? Explain. (b) Is f surjective? Explain.

   (c) Write the function using two-line notation.

9. **Consider the function $f : \mathbb{N} \to \mathbb{N}$ given recursively by $f(0) = 1$ and $f(n + 1) = 2 \cdot f(n)$. Find $f(10)$.**

10. **Suppose $f: \mathbb{N} \to \mathbb{N}$ satisfies the recurrence $f(n + 1) = f(n) + 3$. Note that this is not enough information to define the function, since we don't have an initial condition. For each of the initial conditions below, find the value of $f(5)$.**

(a) f (0) = 0. (b) f (0) = 1.

(c) f (0) = 2. (d) f (0) = 100.

11. Suppose $f: \mathbb{N} \to \mathbb{N}$ satisfies the recurrence relation

$$f(n+1) = \begin{cases} \frac{f(n)}{2} & \text{if } f(n) \text{ is even} \\ 3f(n)+1 & \text{if } f(n) \text{ is odd} \end{cases}.$$

Note that with the initial condition $f(0) = 1$, the values of the function are: $f(1) = 4$, $f(2) = 2$, $f(3) = 1$, $f(4) = 4$, and so on, the images cycling through those three numbers. Thus f is NOT injective (and also certainly not surjective). Might it be under other initial conditions?[3]

(a) If f satisfies the initial condition $f(0) = 5$, is f injective? Explain why or give a specific example of two elements from the domain with the same image.

(b) If f satisfies the initial condition $f(0) = 3$, is f injective? Explain why or give a specific example of two elements from the domain with the same image.

(c) If f satisfies the initial condition $f(0) = 27$, then it turns out that $f(105) = 10$ and no two numbers less than 105 have the same image. Could f be injective? Explain.

(d) Provethatnomatterwhatinitialconditionyouchoose,thefunctioncannotbe surjective.

12. For each function given below, determine whether or not the function is injective and whether or not the function is surjective.

(a) $f: \mathbb{N} \to \mathbb{N}$ given by $f(n) = n + 4$.

(b) $f: \mathbb{Z} \to \mathbb{Z}$ given by $f(n) = n + 4$.

(c) $f: \mathbb{Z} \to \mathbb{Z}$ given by $f(n) = 5n - 8$.

(d) $f: \mathbb{Z} \to \mathbb{Z}$ given by $f(n) = \begin{cases} n/2 & \text{if } n \text{ is even} \\ (n+1)/2 & \text{if } n \text{ is odd.} \end{cases}$

13. Let $A = \{1, 2, 3, \ldots, 10\}$. Consider the function $f: \Pi(A) \to \mathbb{N}$ given by $f(B) = |B|$. That is, f takes a subset of A as an input and outputs the cardinality of that set.

(a) Is f injective? Prove your answer.

(b) Is f surjective? Prove your answer.

3. It turns out this is a really hard question to answer in general. The Collatz conjecture is that no matter what the initial condition is, the function will eventually produce 1 as an output. This is an open problem in mathematics: nobody knows the answer.

(c) Find $f^{-1}(1)$.

(d) Find $f^{-1}(0)$.

(e) Find $f^{-1}(12)$.

14. Let X = $\{n \in \mathbb{N} : 0 \leq n \leq 999\}$ be the set of all numbers with three or fewer digits. Define the function $f : X \rightarrow \mathbb{N}$ by $f(abc) = a + b + c$, where a, b, and c are the digits of the number in X (write numbers less than 100 with leading 0's to make them three digits). For example, $f(253) = 2 + 5 + 3 = 10$.

(a) Let $A = \{n \in X : 113 \leq n \leq 122\}$. Find $f(A)$.

(b) Find $f^{-1}(\{1, 2\})$

(c) Find $f^{-1}(3)$.

(d) Find $f^{-1}(28)$.

(e) Is f injective? Explain.

(f) Is f surjective? Explain.

15. Consider the set $\mathbb{N}^2 = \mathbb{N} \times \mathbb{N}$, the set of all ordered pairs (a, b) where a and b are natural numbers. Consider a function $f : \mathbb{N}^2 \rightarrow \mathbb{N}$ given by $f((a, b)) = a + b$.

(a) Let $A = \{(a, b) \in \mathbb{N}^2 : a, b \leq 10\}$. Find $f(A)$.

(b) Find $f^{-1}(3)$ and $f^{-1}(\{0, 1, 2, 3\})$.

(c) Give geometric descriptions of $f^{-1}(n)$ and $f^{-1}(\{0, 1, \ldots, n\})$ for any $n \geq 1$.

(d) Find $f^{-1}(8)$ and f -1({0, 1, . . . , 8}) .

16. Let $f : X \rightarrow Y$ be some function. Suppose $3 \in Y$. What can you say about $f^{-1}(3)$ if you know,

(a) f is injective? Explain.

(b) f is surjective? Explain.

(c) f is bijective? Explain.

17. Find a set X and a function $f : X \rightarrow \mathbb{N}$ so that $f^{-1}(0) \cup f^{-1}(1) = X$.

18. What can you deduce about the sets X and Y if you know,

(a) there is an injective function $f : X \rightarrow Y$? Explain.

(b) there is a surjective function $f : X \rightarrow Y$? Explain.

(c) there is a bijective function $f : X \rightarrow Y$? Explain.

19. Suppose $f : X \rightarrow Y$ is a function. Which of the following are possible? Explain.

(a) f is injective but not surjective.

(b) f is surjective but not injective.

(c) $|X| = |Y|$ and f is injective but not surjective.

(d) $|X| = |Y|$ and f is surjective but not injective.

(e) $|X| = |Y|$, X and Y are finite, and f is injective but not surjective.

(f) $|X| = |Y|$, X and Y are finite, and f is surjective but not injective.

20. Let $f: X \to Y$ and $g: Y \to Z$ be functions. We can define the composition of f and g to be the function $g f: X \to Z$ for which the image of each $x \in X$ is $g(f(x))$. That is, plug x into f, then plug the result into g (just like composition in algebra and calculus).

(a) If f and g are both injective, must $g \circ f$ be injective? Explain.

(b) If f and g are both surjective, must $g \circ f$ be surjective? Explain.

(c) Suppose $g \circ f$ is injective. What, if anything, can you say about f and g? Explain.

(d) Suppose $g \circ f$ is surjective. What, if anything, can you say about f and g? Explain.

21. Consider the function $f: \mathbb{Z} \to \mathbb{Z}$ given

$$f(n) = \begin{cases} n+1 & \text{if } n \text{ is even} \\ n-3 & \text{if } n \text{ is odd.} \end{cases}$$

(a) Is f injective? Prove your answer.

(b) Is f surjective? Prove your answer.

22. At the end of the semester a teacher assigns letter grades to each of her students. Is this a function? If so, what sets make up the domain and codomain, and is the function injective, surjective, bijective, or neither?

23. In the game of Hearts, four players are each dealt 13 cards from a deck of 52. Is this a function? If so, what sets make up the domain and codomain, and is the function injective, surjective, bijective, or neither?

24. Seven players are playing 5-card stud. Each player initially receives 5 cards from a deck of 52. Is this a function? If so, what sets make up the domain and codomain, and is the function injective, surjective, bijective, or neither?

25. Consider the function $f: N \to N$ that gives the number of handshakes that take place in a room of n people assuming everyone shakes hands with everyone else. Give a recursive definition for this function.

26. Let $f: X \to Y$ be a function and $A \subseteq X$ be a finite subset of the domain. What can you say about the relationship between $|A|$ and $|f(A)|$? Consider both the general case and what happens when you know f is injective, surjective, or bijective.

27. Let $f: X \to Y$ be a function and $B \subseteq Y$ be a finite subset of the c1 odoma1 in. What can you say about the relationship between $|B|$ and $|f^{-1}(B)|$? Consider both the general case and what happens when you know f is injective, surjective, or bijective.

28. Let $f: X \rightarrow Y$ be a function, $A \subseteq X$ and $B \subseteq Y$.

(a) Is $f^{-1}(f(A)=A$? Always, sometimes, never? Explain.

(b) Is $f(f^{-1}(B)) = B$? Always, sometimes, never? Explain.

(c) If one or both of the above do not always hold, is there something else you can say? Will equality always hold for particular types of functions? Is there some other relationship other than equality that would always hold? Explore.

29. Let $f: X \rightarrow Y$ be a function and $A, B \subseteq X$ be subsets of the domain.

(a) Is $f(A \cup B) = f(A) \cup f(B)$? Always, sometimes, or never? Explain.

(b) Is $f(A \cap B) = f(A) \cap f(B)$? Always, sometimes, or never? Explain.

30. Let $f: X \rightarrow Y$ be a function and $A, B \subseteq Y$ be subsets of the codomain.

(a) Is $f^{-1}(A \cup B) = f^{-1}(A) \cup f^{-1}(B)$? Always, sometimes, or never? Explain.

(b) Is $f^{-1}(A \cap B) = f^{-1}(A) \cap f^{-1}(B)$? Always, sometimes, or never? Explain.

CHAPTER-2

COUNTING

INTRODUCTION

Counting is a fundamental concept in mathematics that forms the bedrock upon which many mathematical ideas are built. From basic arithmetic operations to complex algebraic equations and statistical analyses, counting provides the essential framework for quantifying and understanding the numerical relationships that govern our world. Whether tallying discrete objects or measuring continuous quantities, the ability to count accurately and efficiently underpins countless mathematical and scientific endeavors. In this discussion, we explore the various facets of counting, its historical significance, its applications across different fields, and its role in shaping our understanding of the universe through numbers.

One of the first things you learn in mathematics is how to count. Now we want to count large collections of things quickly and precisely. For example:

- In a group of 10 people, if everyone shakes hands with everyone else exactly once, how many handshakes took place?
- How many ways can you distribute 10 girl scout cookies to 7 boy scouts?
- How many anagrams are there of "anagram"?
- Before tackling questions like these, let's look at the basics of counting.

2.1 ADDITIVE AND MULTIPLICATIVE PRINCIPLES

Investigate!

A restaurant offers 8 appetizers and 14 entrées. How many choices do you have if:

- you will eat one dish, either an appetizer or an entrée?
- you are extra hungry and want to eat both an appetizer and an entrée?

Think about the methods you used to solve question 1. Write down the rules for these methods.

Do your rules work? A standard deck of playing cards has 26 red cards and 12 face cards.

- How many ways can you select a card which is either red or a face card?
- How many ways can you select a card which is both red and a face card?
- How many ways can you select two cards so that the first one is red and the second one is a face card?

Attempt the above activity before proceeding

Consider this rather simple counting problem: at Red Dogs and Donuts, there are 14 varieties of donuts, and 16 types of hot dogs. If you want either a donut or a dog, how many options do you have? This isn't too hard, just add 14 and 16. Will that always work? What is important here?

Additive Principle

The **additive principle** states that if event A can occur in m ways, and event *B* can occur in n disjoint ways, then the event "*A* or *B*" can occur in $m + n$ ways.

It is important that the events be **disjoint:** i.e., that there is no way for A and B to both happen at the same time. For example, a standard deck of 52 cards contains 26 red cards and 12 face cards. However, the number of ways to select a card which is either red or a face card is not 26 + 12 = 38.

This is because there are 6 cards which are both red and face cards.

Example 2.1.1

How many two letter "words" start with either A or B? (A **word** is just a string of letters; it doesn't have to be English, or even pronounceable.)

Solution

First, how many two letter words start with A? We just need to select the second letter, which can be accomplished in 26 ways. So there are 26 words starting with A. There are also 26 words that start with B. To select a word which starts with either A or B, we can pick the word from the first 26 or the second 26, for a total of 52 words.

The additive principle also works with more than two events. Say, in addition to your 14 choices for donuts and 16 for dogs, you would also consider eating one of 15 waffles? How many choices do you have now? You would have 14 + 16 + 15 = 45 options.

Example 2.1.2

How many two letter words start with one of the 5 vowels?

Solution.

There are 26 two letter words starting with A, another 26 starting with E, and so on. We will have 5 groups of 26. So we add 26 to itself 5 times. Of course it would be easier to just multiply $5 \cdot 26$. We are really using the additive principle again, just using multiplication as a shortcut.

Example 2.1.3

Suppose you are going for some fro-yo. You can pick one of 6 yogurt choices, and one of 4 toppings. How many choices do you have?

Solution

Break your choices up into disjoint events: A are the choices with the first topping, B the choices featuring the second topping, and so on. There are four events; each can occur in 6 ways (one for each yogurt flavor). The events are disjoint, so the total number of choices is $6 + 6 + 6 + 6 = 24$.

Note that in both of the previous examples, when using the additive principle on a bunch of events all the same size, it is quicker to multiply. This really is the same, and not just because $6 + 6 + 6 + 6 = 4\ 6$. We can first select the topping in 4 ways (that is, we first select which of the disjoint events we will take). For each of those first 4 choices, we now have 6 choices of yogurt. We have:

Multiplicative Principle

The **multiplicative principle** states that if event A can occur in m ways, and each possibility for A allows for exactly n ways for event B, then the event "A and B" can occur in $m \cdot n$ ways.

The multiplicative principle generalizes to more than two events as well.

Example 2.1.4

How many license plates can you make out of three letters followed by three numerical digits?

Solution

Here we have six events: the first letter, the second letter, the third letter, the first digit, the second digit, and the third digit. The first three events can each happen in 26 ways; the last three can each happen in 10 ways. So the total number of license plates will be 26 26 26 10 10 10, using the multiplicative principle.

Does this make sense? Think about how we would pick a license plate. How many choices we would have? First, we need to pick the first letter. There are 26 choices. Now for each of those, there are 26 choices for the second letter: 26 second letters with first letter A, 26 second letters with first letter B, and so on. We add 26 to itself 26 times. Or quicker: there are 26 26 choices for the first two letters.

Now for each choice of the first two letters, we have 26 choices for the third letter. That is, 26 third letters for the first two letters AA, 26 choices for the third letter after starting AB, and so on. There are $26 \cdot 26$ of these 26 third letter choices, for a total of $(26 \cdot 26) \cdot 26$ choices of letters, we have a bunch of choices for the remaining digits.

In fact, there are going to be exactly 1000 choices for the numbers. We can see this because there are 1000 three-digit numbers (000 through 999). This is 10 choices for the first digit, 10 for the second, and 10 for the third. The multiplicative principle says we multiply: $10 \cdot 10 \cdot 10 = 1000$.

All together, there were 263 choices for the three letters, and 10^3 choices for the first three letters, and 10^3 choices for the numbers, so we have a total of $26^3 \cdot 10^3$ choices of license plates.

Careful: "and" doesn't mean "times." For example, how many playing cards are both red and a face card? Not 26. 12. The answer is 6, and we needed to know something about cards to answer that question.

Another caution: how many ways can you select two cards, so that the first one is a red card and the second one is a face card? This looks more like the multiplicative principle (you are counting two separate events) but the answer is not 26. 12 here either. The problem is that while there are 26 ways for the first card to be selected, it is not the case that *for each* of those there are 12 ways to select the second card. If the first card was both red and a face card, then there would be only 11 choices for the second card.

Example 2.1.5 Counting functions

How many functions f: $\{1, 2, 3, 4, 5\} \to \{a, b, c, d\}$ are there?

Solution

Remember that a function sends each element of the domain to exactly one element of the codomain. To determine a function, we just need to specify the image of each element in the domain. Where can we send 1? There are 4 choices. Where can we send 2? Again, 4 choices. What we have here is 5 "events" (picking the image of an element in the domain) each of which can happen in 4 ways (the choices for that image). Thus there are $4. 4. 4. 4. 4 = 4^5$ functions.

This is more than just an example of how we can use the multiplicative principle in a particular counting question. What we have here is a general interpretation of certain applications of the multiplicative principle using rigorously defined mathematical objects: functions. Whenever we have a counting question that asks for the number of outcomes of a repeated event, we can interpret that as asking for the number of functions from $\{1, 2, \ldots, n\}$ (where n is the number of times the event is repeated) to $\{1, 2, \ldots, k\}$ (where k is the number of ways that event can occur).

Counting With Sets

Do you believe the additive and multiplicative principles? How would you convince someone they are correct? This is surprisingly difficult. They seem so simple, so obvious. But why do they work?

To make things clearer, and more mathematically rigorous, we will use sets. Do not skip this section! It might seem like we are just trying to give a proof of these principles, but we are doing a lot more. If we understand the additive and multiplicative principles rigorously, we will be better at applying them, and knowing when and when not to apply them at all.

We will look at the additive and multiplicative principles in a slightly different way. Instead of thinking about event A and event B, we want to think of a set A and a set B. The sets will contain all the different ways the event can happen. (It will be helpful to be able to switch back and forth between these two models when checking that we have counted correctly.) Here's what we mean:

Example 2.1.6

Suppose you own 9 shirts and 5 pairs of pants.

- How many outfits can you make?
- If today is half-naked-day, and you will wear only a shirt or only a pair of pants, how many choices do you have?

Solution

By now you should agree that the answer to the first question is $9 \cdot 5 = 45$ and the answer to the second question is $9 + 5 = 14$. These are the multiplicative and additive principles. There are two events: picking a shirt and picking a pair of pants.

The first event can happen in 9 ways and the second event can happen in 5 ways. To get both a shirt and a pair of pants, you multiply. To get just one article of clothing, you add.

Now look at this using sets. There are two sets, call them S and P. The set S contains all 9 shirts so $|S| = 9$ while $|P| = 5$, since there are 5 elements in the set P (namely your 5 pairs of pants). What are we asking in terms of these sets? Well in question 2, we really want $|S \cup P|$, the number of elements in the union of shirts and pants.

This is just $|S|+|P|$ (since there is no overlap; $|S \cap P| = 0$). Question 1 is slightly more complicated. Your first guess might be to find $|S \cap P|$, but this is not right (there is nothing in the intersection). We are not asking for how many clothing items are both a shirt and a pair of pants. Instead, we want one of each. We could think of this as asking how many pairs (x, y) there are, where x is a shirt and y is a pair of pants. As we will soon verify, this number is $|S| \cdot |P|$.

From this example we can see right away how to rephrase our additive principle in terms of sets:

Additive Principle (with sets)

Given two sets A and B, if $A \cap B = \emptyset$ (that is, if there is no element in common to both A and B), then

$$|A \cup B| = |A| + |B|.$$

This hardly needs a proof. To find $A \cup B$, you take everything in A and throw in everything in B. Since there is no element in both sets already, you will have $|A|$ things and add $|B|$ new things to it. This is what adding does! Of course, we can easily extend this to any number of disjoint sets. From the example above, we see that in order to investigate the multiplicative principle carefully, we need to consider ordered pairs. We should define this carefully:

Cartesian Product

Given sets A and B, we can form the *set* to be the set of all ordered pairs (x, y) where x is an element of A and y is an element of B. We call A × B the **Cartesian product** of A and B.

$$A \times B = \{(x, y) : x \in A \wedge y \in B\}$$

Example 2.1.7

Let A = {1, 2} and B = {3, 4, 5}. Find $A \times B$.

Solution

We want to find ordered pairs (a, b) where a can be either 1 or 2 and b can be either 3, 4, or 5. A × B is the set of all of these pairs:

$$A \times B = \{(1, 3), (1, 4), (1, 5), (2, 3), (2, 4), (2, 5)\}.$$

The question is, what is $|A \times B|$? To figure this out, write out $A \times B$. $A = \{a_1, a_2, a_3, \ldots, a_m\}$ and $B = \{b_1, b_2, b_3, \ldots, b_n\}$ (so $|A| = m$ and $|B| = n$). The set A × B contains all pairs with the first half of the pair being some $a_i \in A$ and the second being one of the $b_j \in B$. In other words:

$$A \times B = \{(a_1, b_1), (a_1, b_2), (a_1, b_3), \ldots (a_1, b_n),$$
$$(a_2, b_1), (a_2, b_2), (a_2, b_3), \ldots (a_2, b_n),$$
$$(a_3, b_1), (a_3, b_2), (a_3, b_3), \ldots (a_3, b_n),$$

$$\vdots$$

$$(a_m, b_1), (a_m, b_2), (a_m, b_3), \ldots (a_m; b_n)\}.$$

Notice what we have done here: we made m rows of n pairs, for a total of $m. n$ pairs.

Each row above is really $\{a_i\} \times B$ for some $a_i \in A$. That is, we fixed the A-element. Broken up this way, we have

$$A \times B = (\{a_1\} \times B) \cup (\{a_2\} \times B) \cup (\{a_3\} \times B) \cup \cdots \cup (\{a_m\} \times B).$$

So $A \times B$ is really the union of m disjoint sets. Each of those sets has n elements in them. The total (using the additive principle) is $n + n + n + \cdots + n = m . n$.

To summarize:

Multiplicative Principle (with sets)

Given two sets A and B, we have $|A \times B| = |A| \cdot |B|$.

Again, we can easily extend this to any number of sets.

Investigate!

A recent buzz marketing campaign for Village Inn surveyed patrons on their pie preferences. People were asked whether they enjoyed (A) Apple, (B) Blueberry or (C) Cherry pie (respondents answered yes or no to each type of pie, and could say yes to more than one type). The following table shows the results of the survey.

Pies enjoyed:	A	B	C	AB	AC	BC	ABC
Number of people:	20	13	26	9	15	7	5

How many of those asked enjoy at least one of the kinds of pie? Also, explain why the answer is not 95.

Attempt the above activity before proceeding

While we are thinking about sets, consider what happens to the additive principle when the sets are NOT disjoint. Suppose we want to find A B and know that $|A| = 10$ and $|B| = 8$. This is not enough information though. We do not know how many of the 8 elements in B are also elements of A. However, if we also know that $|A \cap B| = 6$, then we can say exactly how many elements are in A, and, of those, how many are in B and how many are not (6 of the 10 elements are in B, so 4 are in A but not in B). We could fill in a Venn diagram as follows:

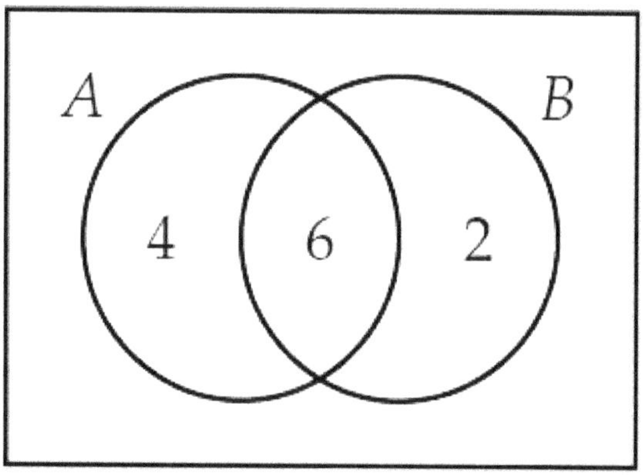

This says there are 6 elements in A∩B, 4 elements in A\B and 2 elements in B\A. Now these three sets are disjoint, so we can use the additive principle to find the number of elements in A∪B. It is 6+4+2 = 12. This will always work, but drawing a Venn diagram is more than we need to do. In fact, it would be nice to relate this problem to the case where A and B are disjoint. Is there one rule we can make that works in either case?

Here is another way to get the answer to the problem above. Start by just adding $|A| + |B|$. This is 10 + 8 = 18, which would be the answer if $|A \cap B| = 0$. We see that we are off by exactly 6, which just so happens to be $|A \cap B|$. So perhaps we guess,

$$|A \cup B| = |A| + |B| - |A \cap B|.$$

This works for this one example. Will it always work? Think about what we are doing here. We want to know how many things are either in A or B (or both). We can throw in everything in A, and everything in B.

This would give $|A| + |B|$ many elements. But of course when you actually take the union, you do not repeat elements that are in both. So far we have counted every element in A ∩ B exactly twice: once when we put in the elements from A and once when we included the elements from B. We correct by subtracting out the number of elements we have counted twice. So we added them in twice, subtracted once, leaving them counted only one time.

In other words, we have:

Cardinality of a union (2 sets)

For any finite sets *A* and *B*,

$$|A \cup B| = |A| + |B| - |A \cap B|.$$

We can do something similar with three sets.

Example 2.1.8

An examination in three subjects, Algebra, Biology, and Chemistry, was taken by 41 students. The following table shows how many students failed in each single subject and in their various combinations:

Subject:	A	B	C	AB	AC	BC	ABC
Failed:	12	5	8	2	6	3	1

How many students failed at least one subject?

Solution

The answer is not 37, even though the sum of the numbers above is 37. For example, while 12 students failed Algebra, 2 of those students also failed Biology, 6 also failed Chemistry, and 1 of those failed all three subjects. In fact, that 1 student who failed all three subjects is counted a total of 7 times in the total 37. To clarify things, let us think of the students who failed Algebra as the elements of the set A, and similarly for sets B and C. The one student who failed all three subjects is the lone element of the set A ∩ B ∩ C. Thus, in Venn diagrams:

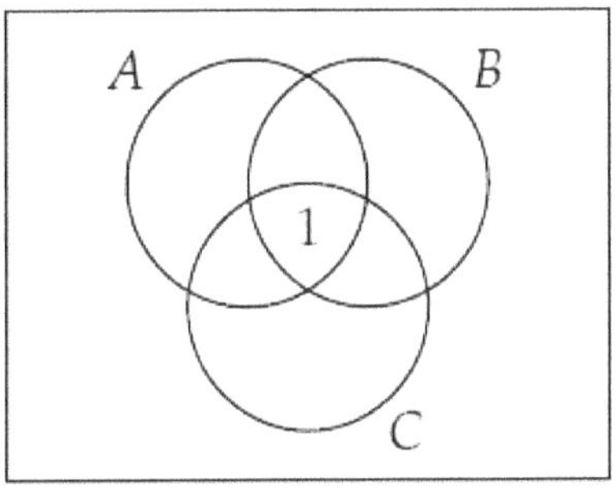

Now let's fill in the other intersections. We know A B contains 2 elements, but 1 element has already been counted. So we should put a 1 in the region where A and B intersect (but C does not). Similarly, we calculate the cardinality of (A ∩ C) \ B, and (B ∩ C) \ A:

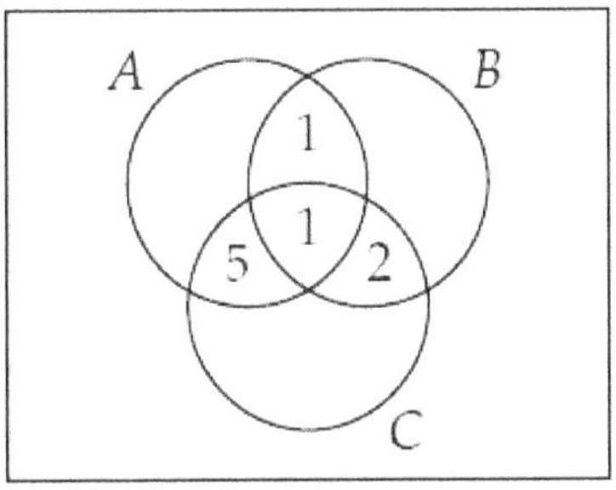

Next, we determine the numbers which should go in the remaining regions, including outside of all three circles. This last number is the number of students who did not fail any subject:

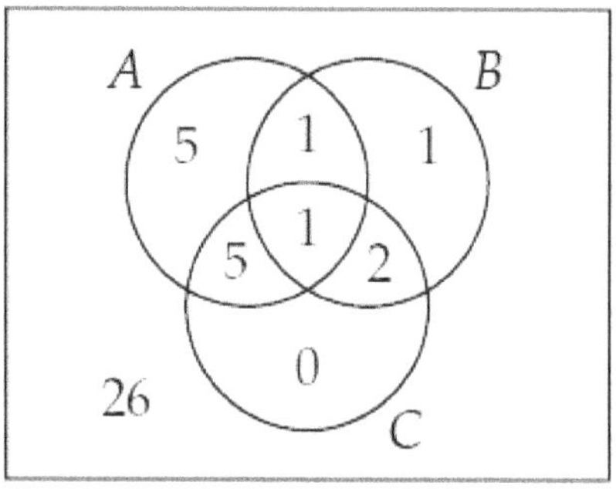

We found 5 goes in the "*A* only" region because the entire circle for A needed to have a total of 12, and 7 were already accounted for. Similarly, we calculate the "*B* only" region to contain only 1 student and the "*C* only" region to contain no students.

Thus the number of students who failed at least one class is 15 (the sum of the numbers in each of the eight disjoint regions). The number of students who passed all three classes is 26: the total number of students, 41, less the 15 who failed at least one class.

Note that we can also answer other questions. For example, how many students failed just Chemistry? None. How many passed Algebra but failed both Biology and Chemistry? This corresponds to the region inside both B and C but outside of A, containing 2 students.

Could we have solved the problem above in an algebraic way? While the additive principle generalizes to any number of sets, when we add a third set here, we must be careful. With two sets, we needed to know the cardinalities of A, B, and A B in order to find the cardinality of A B. With three sets we need more information. There are more ways the sets can combine. Not surprisingly then, the formula for cardinality of the union of three non-disjoint sets is more complicated:

Cardinality of a union (3 sets)

For any finite sets *A*, *B*, and *C*,

$$|A \cup B \cup C| \quad |A|+|B|+|C|-|A \cap B|-|A \cap C|-|B \cap C|+|A \cap B \cap C|$$

To determine how many elements are in at least one of A, B, or C we add up all the elements in each of those sets. However, when we do that, any element in both A and B is counted twice. Also, each element in both A and C is counted twice, as are elements in B and C, so we take each of those out of our sum once. But now what about the elements which are in $A \cap B \cap C$ (in all three sets)? We added them in three times, but also removed them three times. They have not yet been counted. Thus we add those elements back in at the end.

Returning to our example above, we have $|A| = 12$, $|B| = 5$, $|C| = 8$.

We also have $|A \cap B| = 2$, $|A \cap C| = 6$, $|B \cap C| = 3$, and $|A \cap B \cap C| = 1$.

Therefore:

$$|A \cup B \cup C| = 12 + 5 + 8 - 2 - 6 - 3 + 1 = 15.$$

This is what we got when we solved the problem using Venn diagrams.

This process of adding in, then taking out, then adding back in, and so on is called the *Principle of Inclusion/Exclusion,* or simply PIE. We will return to this counting technique later to solve for more complicated problems (involving more than 3 sets).

Exercises

1. **Your wardrobe consists of 5 shirts, 3 pairs of pants, and 17 bow ties. bow ties How many different outfits can you make?**

2. **For your college interview, you must wear a tie. You own 3 regular (boring) ties and 5 (cool) bow ties.**

   (a) How many choices do you have for your neck-wear?

   (b) You realize that the interview is for clown college, so you should probably wear both a regular tie and a bow tie. How many choices do you have now?

   (c) For the rest of your outfit, you have 5 shirts, 4 skirts, 3 pants, and 7 dresses. You want to select either a shirt to wear with a skirt or pants, or just a dress. How many outfits do you have to choose from?

3. **Your Blu-ray collection consists of 9 comedies and 7 horror movies. Give an example of a question for which the answer is:**

   (a) 16. (b) 63.

4. **hexadecimal We usually write numbers in decimal form (or base 10), meaning numbers are composed using 10 different "digits" {0, 1, . . . , 9] . Sometimes though it is useful to write numbers hexadecimal or base 16. Now there are 16 distinct digits that can be used to form numbers: {0, 1, . . . , 9, A, B, C, D, E, F} . So for example, a 3 digit hexadecimal number might be 2B8.**

   (a) How many 2-digit hexadecimals are there in which the first digit is E or F? Explain your answer in terms of the additive principle (using either events or sets).

   (b) Explain why your answer to the previous part is correct in terms of the multiplicative principle (using either events or sets). Why do both the additive and multiplicative principles give you the same answer?

   (c) How many 3-digit hexadecimals start with a letter (A-F) and end with a numeral (0-9)? Explain.

(d) How many 3-digit hexadecimals start with a letter (A-F) or end with a numeral (0-9) (or both)? Explain.

5. **Suppose you have sets A and B with $|A| = 10$ and $|B| = 15$.**

   (a) What is the largest possible value for $|A \cap B|$?

   (b) What is the smallest possible value for $|A \cap B|$?

   (c) What are the possible values for $|A \cup B|$?

6. **If $|A| = 8$ and $|B| = 5$, what is $|A \cup B| + |A \cap B|$?**

7. **A group of college students were asked about their TV watching habits. Of those surveyed, 28 students watch *The Walking Dead,* 19 watch *The Blacklist,* and 24 watch *Game of Thrones.* Additionally, 16 watch *The Walking Dead* and *The Blacklist,* 14 watch *The Walking Dead* and *Game of Thrones,* and 10 watch *The Blacklist* and *Game of Thrones.* There are 8 students who watch all three shows. How many students surveyed watched at least one of the shows?**

8. **In a recent survey, 30 students reported whether they liked their potatoes Mashed, French-fried, or Twice-baked. 15 liked them mashed, 20 liked French fries, and 9 liked twice baked potatoes. Additionally, 12 students liked both mashed and fried potatoes, 5 liked French fries and twice baked potatoes, 6 liked mashed and baked, and 3 liked all three styles. How many students hate potatoes? Explain why your answer is correct.**

9. **For how many $n \in \{1, 2, \ldots, 500\}$ is n a multiple of one or more of 5, 6, or 7?**

10. **How many positive integers less than 1000 are multiples of 3, 5, or 7? Explain your answer using the Principle of Inclusion/Exclusion.**

11. **Let A, B, and C be sets.**

    (a) Find $|(A \cup C) \setminus B|$ provided $|A| = 50$, $|B| = 45$, $|C| = 40$, $|A \cap B| = 20$, $|A \cap C| = 15$, $|B \cap C| = 23$, and $|A \cap B \cap C| = 12$.

    (b) Describe a set in terms of A, B, and C with cardinality 26.

12. **Consider all 5 letter "words" made from the letters a through h. (Recall, words are just strings of letters, not necessarily actual English words.)**

    (a) How many of these words are there total?

    (b) How many of these words contain no repeated letters?

    (c) How many of these words start with the sub-word "aha"?

    (d) How many of these words either start with "aha" or end with "bah" or both?

    (e) How many of the words containing no repeats also do not contain the sub-word "bad"?

13. **For how many three digit numbers (100 to 999) is the *sum of the digits* even? (For example, 343 has an even sum of digits: 3 + 4 + 3 = 10 which is even.) Find the answer and explain why it is correct in at least two *different* ways.**

14. The number 735000 factors as $2^3 . 3 . 5^4 7^2$. How many divisors does it have? Explain your answer using the multiplicative principle.

2.2 BINOMIAL COEFFICIENTS

Investigate!

In chess, a rook can move only in straight lines (not diagonally). Fill in each square of the chess board below with the number of different shortest paths the rook, in the upper left corner, can take to get to that square. For example, one square is already filled in. There are six different paths from the rook to the square: DDRR (down down right right), DRDR, DRRD, RDDR, RDRD and RRDD.

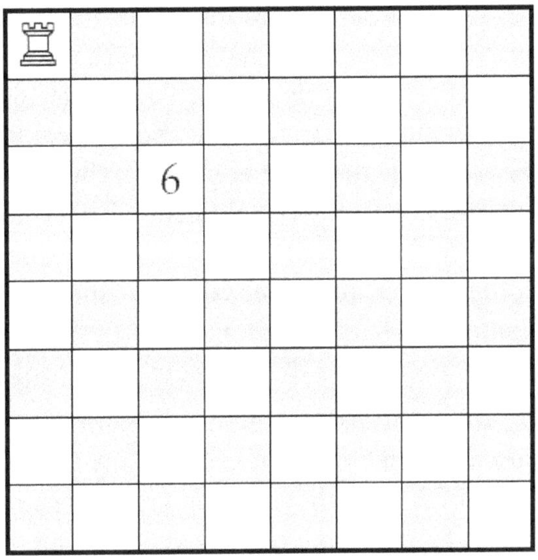

Attempt the above activity before proceeding

Here are some apparently different discrete objects we can count: subsets, bit strings, lattice paths, and binomial coefficients. We will give an example of each type of counting problem (and say what these things even are). As we will see, these counting problems are surprisingly similar.

Subsets

Subsets should be familiar, otherwise read over Section 0.3 again. Suppose we look at the set $A = \{1, 2, 3, 4, 5\}$. How many subsets of A contain exactly 3 elements?

First, a simpler question: How many subsets of A are there total? In other words, what is $|\Pi(A)|$ (the cardinality of the power set of A)? Think about how we would build a subset. We need to decide, for each of the elements of A, whether or not to include the element in our subset. So we need to decide "yes" or "no" for the element 1. And for each choice we make, we need to decide "yes" or "no" for the element 2. And so on.

For each of the 5 elements, we have 2 choices. Therefore the number of subsets is simply $2 \cdot 2 \cdot 2 \cdot 2 \cdot 2 = 2^5$ (by the multiplicative principle).

Of those 32 subsets, how many have 3 elements? This is not obvious. Note that we cannot just use the multiplicative principle. Maybe we want to say we have 2 choices (yes/no) for the first element, 2 choices for the second, 2 choices for the third, and then only 1 choice for the other two. But what if we said "no" to one of the first three elements? Then we would have two choices for the 4th element. What a mess!

Another (bad) idea: we need to pick three elements to be in our subset. There are 5 elements to choose from. So there are 5 choices for the first element, and for each of those 4 choices for the second, and then 3 for the third (last) element. The multiplicative principle would say then that there are a total of 5. 4. 3 = 60 ways to select the 3-element subset. But this cannot be correct (60 > 32 for one thing). One of the outcomes we would get from these choices would be the set $\{3, 2, 5\}$, by choosing the element 3 first, then the element 2, then the element 5. Another outcome would be $\{5, 2, 3\}$ by choosing the element 5 first, then the element 2, then the element 3. But these are the same set! We can correct this by dividing: for each set of three elements, there are 6 outcomes counted among our 60 (since there are 3 choices for which element we list first, 2 for which we list second, and 1 for which we list last). So we expect there to be 10 3-element subsets of A.

Is this right? Well, we could list out all 10 of them, being very systematic in doing so, to make sure we don't miss any or list any twice. Or we could try to count how many subsets of A don't have 3 elements in them. How many have no elements? Just 1 (the empty set). How many have 5? Again, just 1. These are the cases in which we say "no" to all elements, or "yes" to all elements. Okay, what about the subsets which contain a single element? There are 5 of these. We must say "yes" to exactly one element, and there are 5 to choose from. This is also the number of subsets containing 4 elements. Those are the ones for which we must say "no" to exactly one element.

So far we have counted 12 of the 32 subsets. We have not yet counted the subsets with cardinality 2 and with cardinality 3. There are a total of 20 subsets left to split up between these two groups. But the number of each must be the same! If we say "yes" to exactly two elements, that can be accomplished in exactly the same number of ways as the number of ways we can say "no" to exactly two elements. So the number of 2-element subsets is equal to the number of 3-element subsets. Together there are 20 of these subsets, so 10 each.

Number of elements:	0	1	2	3	4	5
Number of subsets:	1	5	10	10	5	1

Bit Strings

"Bit" is short for "binary digit," so a bit string is a string of binary digits. The binary digits are simply the numbers 0 and 1. All of the following are bit strings:

1001	0	1111	1010101010.

The number of bits (0's or 1's) in the string is the **length** of the string; the strings above have lengths 4, 1, 4, and 10 respectively. We also can ask how many of the bits are 1's. The number of 1's in a bit string is the **weight** of the string; the weights of the above strings are 2, 0, 4, and 5 respectively.

Bit Strings

- An *n*-**bit string** is a bit string of length n. That is, it is a string containing n symbols, each of which is a bit, either 0 or 1.
- The **weight** of a bit string is the number of 1's in it.
- $\mathbf{B}^n$ is the set of all n-bit strings.
- $\mathbf{B}^n_k$ is the set of all n-bit strings of weight k.

For example, the elements of the set $\mathbf{B}^3_2$ are the bit strings 011, 101, and 110. Those are the only strings containing three bits exactly two of which are 1's.

The counting questions: How many bit strings have length 5? How many of those have weight 3? In other words, we are asking for the cardinalities $|\mathbf{B}^5|$ and $|\mathbf{B}^5_3|$.

To find the number of 5-bit strings is straight forward. We have 5 bits, and each can either be a 0 or a 1. So there are 2 choices for the first bit, 2 choices for the second, and so on. By the multiplicative principle, there are 2. 2. 2. 2. 2 = 2^5 = 32 such strings.

Finding the number of 5-bit strings of weight 3 is harder. Think about how such a string could start. The first bit must be either a 0 or a 1. In the first case (the string starts with a 0), we must then decide on four more bits. To have a total of three 1's, among those four remaining bits there must be three 1's. To count all of these strings, we must include all 4-bit strings of weight 3. In the second case (the string starts with a 1), we still have four bits to choose, but now only two of them can be 1's, so we should look at all the 4-bit strings of weight 2. So the strings in $\mathbf{B}^5_3$ all have the form $1\mathbf{B}^4_2$ (that is, a 1 followed by a string from $\mathbf{B}^4_2$) or $0\mathbf{B}^4_.$ These two sets are disjoint, so we can use the additive principle:

$$|\mathbf{B}^5_3| = |\mathbf{B}^4_2| + |\mathbf{B}^4_3|.$$

This is an example of a **recurrence relation.** We represented one instance of our counting problem in terms of two simpler instances of the problem. If only we knew the cardinalities of $\mathbf{B}^4_2$ and $\mathbf{B}^4_3$. Repeating the same reasoning,

$$|\mathbf{B}^4_2| = |\mathbf{B}^3_1| + |\mathbf{B}^3_2| \quad \text{and} \quad |\mathbf{B}^4_3| = |\mathbf{B}^3_2| + |\mathbf{B}^3_3|.$$

We can keep going down, but this should be good enough. Both $\mathbf{B}^3_1$ and $\mathbf{B}^3_2$ contain 3 bit strings: we must pick one of the three bits to be a 1 (three ways to do that) or one of the three bits to be a 0 (three ways to do that). Also, $\mathbf{B}^3_3$ contains just one string: 111. Thus $|\mathbf{B}^4_2| = 6$ and $|\mathbf{B}^4_3| = 4$,which puts $\mathbf{B}^5_3$ at a total of 10 strings.

But wait —32 and 10 were the answers to the counting questions about subsets. Coincidence? Not at all. Each bit string can be thought of as a code for a subset. To represent the subsets of $A = \{1, 2, 3, 4, 5\}$, we can use 5-bit strings, one bit for each element of A. Each bit in the string is a 0 if its corresponding element of A is not in the subset, and a 1 if the element of A is in the subset. Remember, deciding the subset amounted to a sequence of five yes/no votes for the elements of A. Instead of yes, we put a 1; instead of no, we put a 0.

For example, the bit string 11001 represents the subset {1, 2, 5} since the first, second and fifth bits are 1's. The subset {3, 5} would be coded by the string 00101. What we really have here is a bijection from $\Pi(A)$ to $\mathbf{B}^5$.

Now for a subset to contain exactly three elements, the corresponding bit string must contain exactly three 1's. In other words, the weight must be 3. Thus counting the number of 3-element subsets of A is the same as counting the number 5-bit strings of weight 3.

Lattice Paths

The **integer lattice** is the set of all points in the Cartesian plane for which both the x and y coordinates are integers. If you like to draw graphs on graph paper, the lattice is the set of all the intersections of the grid lines.

A **lattice path** is one of the shortest possible paths connecting two points on the lattice, moving only horizontally and vertically. For example, here are three possible lattice paths from the point (0, 0) to (3, 2):

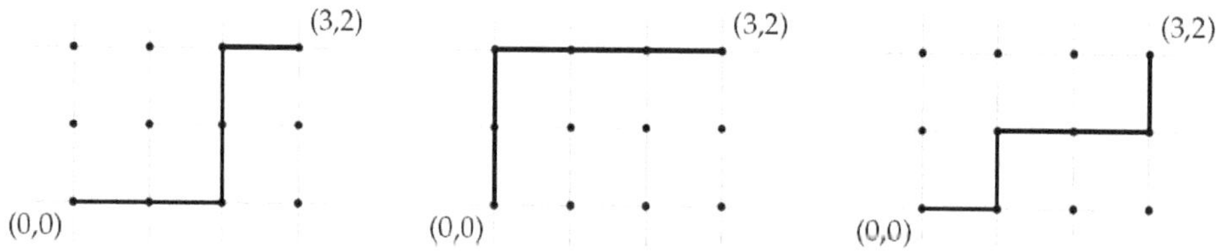

Notice to ensure the path is the *shortest* possible, each move must be either to the right or up. Additionally, in this case, note that no matter what path we take, we must make three steps right and two steps up. No matter what order we make these steps, there will always be 5 steps. Thus each path has *length* 5.

The counting question: how many lattice paths are there between (0, 0) and (3, 2) ? We could try to draw all of these, or instead of drawing them, maybe just list which direction we travel on each of the 5 steps.

One path might be RRUUR, or maybe UURRR, or perhaps RURRU (those correspond to the three paths drawn above). So how many such strings of R's and U's are there?

Notice that each of these strings must contain 5 symbols. Exactly 3 of them must be R's (since our destination is 3 units to the right). This seems awfully familiar. In fact, what if we used 1's instead of R's and 0's instead of U's? Then we would just have 5-bit strings of weight 3. There are 10 of those, so there are 10 lattice paths from (0,0) to (3,2).

The correspondence between bit strings and lattice paths does not stop there. Here is another way to count lattice paths. Consider the lattice shown below:

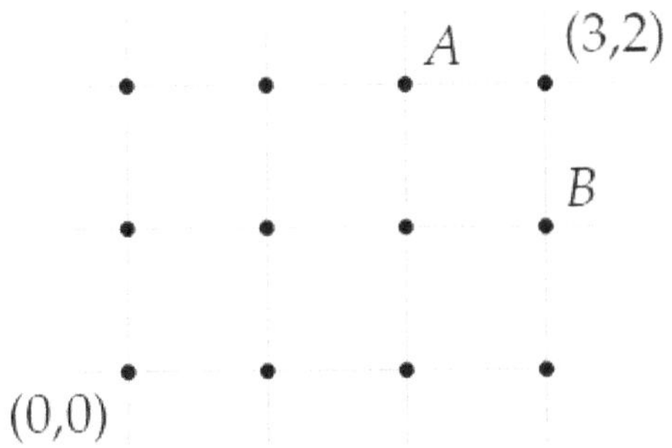

Any lattice path from (0,0) to (3,2) must pass through exactly one of *A* and *B*. The point A is 4 steps away from (0,0) and two of them are towards the right. The number of lattice paths to A is the same as the number of 4-bit strings of weight 2, namely 6. The point B is 4 steps away from (0,0), but now 3 of them are towards the right. So the number of paths to point B is the same as the number of 4-bit strings of weight 3, namely 4. So the total number of paths to (3,2) is just 6 + 4. This is the same way we calculated the number of 5-bit strings of weight 3. The point: the exact same recurrence relation exists for bit strings and for lattice paths.

Binomial Coefficients

Binomial coefficients are the coefficients in the expanded version of a binomial, such as $(x + y)^5$. What happens when we multiply such a binomial out? We will expand $(x + y)^n$ for various values of *n*. Each of these are done by multiplying everything out (i.e., FOIL-ing) and then collecting like terms.

$$(x + y)^1 = x + y$$

$$(x + y)^2 = x^2 + 2xy + y^2$$

$$(x + y)^3 = x^3 + 3x^2 y + 3xy^2 + y^3$$

$$(x + y)^4 = x^4 + 4x^3y + 6x^2 y^2 + 4xy^3 + y^4.$$

In fact, there is a quicker way to expand the above binomials. For example, consider the next one, $(x + y)^5$. What we are really doing is multiplying out,

$$(x + y)(x+y)(x + y)(x+y)(x+y)$$

If that looks daunting, go back to the case of $(x+y)^3 = (x+y)(x+y)(x+y)$. Why do we only have one x3 and y3 but three $x^2 y$ and xy^2 terms? Every time we distribute over an (x + y) we create two copies of what is left, one multiplied by x, the other multiplied by y. To get x^3, we need to pick the "multiplied by x" side every time (we don't have any y's in the term). This will only happen once. On the other hand, to get $x^2 y$ we need to select the x side twice and the y side once. In other words, we need to pick one of the three x + y terms to "contribute" their y.

Similarly, in the expansion of $(x + y)^5$, there will be only one x^5 term and one y^5 term. This is because to get an x^5, we need to use the x term in each of the copies of the binomial (x + y), and similarly for y^5. What about x^4y? To get terms like this, we need to use four x's and one y, so we need exactly one of the five binomials to contribute a y. There are 5 choices for this, so there are 5 ways to get x^4y, so the coefficient of x^4y is 5. This is also the coefficient for xy^4 for the same (but opposite) reason: there are 5 ways to pick which of the 5 binomials contribute the single x.

So far we have

$$(x + y)^5 = x^5 + 5x^4y + \underline{?}\,x^3y^2 + \underline{?}\,x^2y^3 + 5xy^4 + y^5.$$

We still need the coefficients of $x^3 y^2$ and $x^2 y^3$. In both cases, we need to pick exactly 3 of the 5 binomials to contribute one variable, the other two to contribute the other. Wait. This sounds familiar. We have 5 things, each can be one of two things, and we need a total of 3 of one of them. That's just like taking 5 bits and making sure exactly 3 of them are 1's. So the coefficient of $x^3 y^2$ (and also $x^2 y^3$) will be exactly the same as the number of bit strings of length 5 and weight 3, which we found earlier to be 10. So we have:

$$(x + y)^5 = x^5 + 5x^4y + 10x^3y^2 + 10x^2y^3 + 5xy^4 + y^5.$$

These numbers we keep seeing over and over again. They are the number of subsets of a particular size, the number of bit strings of a particular weight, the number of lattice paths, and the coefficients of these binomial products. We will call them binomial coefficients. We even have a special symbol for them: $\binom{n}{k}$.

Binomial Coefficients

For each integer n ≥ 0 and integer k with 0 ≤ k ≤ n there is a number read "*n* choose *k*." We have: $\binom{n}{k}$

- $\binom{n}{k} = |B^n_k|$, the number of n-bit strings of weight k.
- $\binom{n}{k}$ is the number of subsets of a set of size n each with cardinality k.
- $\binom{n}{k}$ is the number of lattice paths of length n containing k steps to the right.
- $\binom{n}{k}$ is the coefficient of $x^k y^{n-k}$ in the expansion of $(x + y)^n$.
- $\binom{n}{k}$ is the number of ways to select k objects from a total of n objects.

The last bullet point is usually taken as the definition of . Out of $\binom{n}{k}$. Out of an objects we must choose k of them, so there are n choose k ways of doing this. Each of our counting problems above can be viewed in this way:

How many subsets of {1, 2, 3, 4, 5} contain exactly 3 elements? We must choose 3 of the 5 elements to be in our subset. There are $\binom{5}{3}$ ways to do this, so there are $\binom{5}{3}$ such subsets.

How many bit strings have length 5 and weight 3? We must choose 3 of the 5 bits to be 1's. There are $\binom{5}{3}$ ways to do this, so there are $\binom{5}{3}$ such bit strings.

How many lattice paths are there from (0,0) to (3,2)? We must choose 3 of the 5 steps to be towards the right. There are $\binom{5}{3}$ ways to do this, so there are $\binom{5}{3}$ such lattice paths.

What is the coefficient of $x^3 y^2$ in the expansion of $(x + y)^5$? We must choose 3 of the 5 copies of the binomial to contribute an x. There are $\binom{5}{3}$ ways to do this, so the coefficient is $\binom{5}{3}$.

It should be clear that in each case above, we have the right answer. All we had to do is phrase the question correctly and it became obvious that $\binom{5}{3}$ is correct. However, this does not tell us that the answer is in fact 10 in each case. We will eventually find a formula for $\binom{n}{k}$, but for now, look back at how we arrived at the answer 10 in our counting problems above. It all came down to bit strings, and we have a recurrence relation for bit strings:

$$|\mathbf{B}^n_k| = |\mathbf{B}^{n-1}_{k-1}| + |\mathbf{B}^{n-1}_k|.$$

Remember, this is because we can start the bit string with either a 1 or a 0. In both cases, we have n^{-1} more bits to pick. The strings starting with 1 must contain k^{-1} more 1's, while the strings starting with 0 still need k more 1's.

Since $|\mathbf{B}^n_k| = \binom{n}{k}$, the same recurrence relation holds for binomial coefficients:

RECURRENCE RELATION FOR $\binom{n}{k}$

$$\binom{n}{k} = \binom{n-1}{k-1} + \binom{n-1}{k}.$$

Pascal's Triangle

Let's arrange the binomial coefficients $\binom{n}{k}$ into a triangle like follows:

$$\begin{array}{ccccccccc}
 & & & & \binom{0}{0} & & & & \\
 & & & \binom{1}{0} & & \binom{1}{1} & & & \\
 & & \binom{2}{0} & & \binom{2}{1} & & \binom{2}{2} & & \\
 & \binom{3}{0} & & \binom{3}{1} & & \binom{3}{2} & & \binom{3}{3} & \\
\binom{4}{0} & & \binom{4}{1} & & \binom{4}{2} & & \binom{4}{3} & & \binom{4}{4}
\end{array}$$

This can continue as far down as we like. The recurrence relation for $\binom{n}{k}$ tells us that each entry in the triangle is the sum of the two entries above it. The entries on the sides of the triangle are always 1. This is because $\binom{n}{0} = 1$ for all n since there is only one way to pick 0 of n objects and $\binom{n}{0} = 1$ since there is one way to select all n out of n objects. Using the recurrence relation, and the fact that the sides of the triangle are 1's, we can easily replace all the entries above with the correct values of $\binom{n}{k}$.

Doing so gives us Pascal's triangle

We can use Pascal's triangle to calculate binomial coefficients. For example, using the triangle below, we can find $\binom{12}{6} = 924$.

Pascal's Triangle

Exercises

1. **Let $S = \{1, 2, 3, 4, 5, 6\}$**
   (a) How many subsets are there total?
   (b) How many subsets have $\{2, 3, 5\}$ as a subset?
   (c) How many subsets contain at least one odd number?
   (d) How many subsets contain exactly one even number?
2. **Let $S = \{1, 2, 3, 4, 5, 6\}$**
   (a) How many subsets are there of cardinality 4?
   (b) How many subsets of cardinality 4 have $\{2, 3, 5\}$ as a subset?
   (c) How many subsets of cardinality 4 contain at least one odd number?
   (d) How many subsets of cardinality 4 contain exactly one even number?
3. **Let $A = \{1, 2, 3, \ldots, 9\}$.**
   (a) How many subsets of A are there? That is, find $|\Pi(A)|$. Explain.
   (b) How many subsets of A contain exactly 5 elements? Explain.
   (c) How many subsets of A contain only even numbers? Explain.
   (d) How many subsets of A contain an even number of elements? Explain.

4. **How many 9-bit strings (that is, bit strings of length 9) are there which:**
   (a) Start with the sub-string 101? Explain.
   (b) Have weight 5 (i.e., contain exactly five 1's) and start with the sub-string 101? Explain.
   (c) Either start with 101 or end with 11 (or both)? Explain.
   (d) Have weight 5 and either start with 101 or end with 11 (or both)? Explain.
5. **You break your piggy-bank to discover lots of pennies and nickels. You start arranging these in rows of 6 coins.**
   (a) You find yourself making rows containing an equal number of pennies and nickels. For fun, you decide to lay out every possible such row. How many coins will you need?
   (b) How many coins would you need to make all possible rows of 6 coins (not necessarily with equal number of pennies and nickels)?
6. **How many 10-bit strings contain 6 or more 1's?**
7. **How many subsets of $\{0, 1, \ldots, 9\}$ have cardinality 6 or more?**
8. **What is the coefficient of x^{12} in $(x + 2)^{15}$?**
9. **What is the coefficient of x^9 in the expansion of $(x + 1)^{14} + x^3(x + 2)^{15}$?**
10. **How many lattice paths start at (3,3) and**
    (a) end at (10,10)?
    (b) end at (10,10) and pass through (5,7)?
    (c) end at (10,10) and avoid (5,7)?
11. **Gridtown USA, besides having excellent donut shops, is known for its precisely laid out grid of streets and avenues. Streets run east-west, and avenues north-south, for the entire stretch of the town, never curving and never interrupted by parks or schools or the like. Suppose you live on the corner of 3rd and 3rd and work on the corner of 12th and 12th. Thus you must travel 18 blocks to get to work as quickly as possible.**
    (a) How many different routes can you take to work, assuming you want to get there as quickly as possible? Explain.
    (b) Now suppose you want to stop and get a donut on the way to work, from your favorite donut shop on the corner of 10th ave and 8th st. How many routes to work, stopping at the donut shop, can you take (again, ensuring the shortest possible route)? Explain.
    (c) Disaster Strikes Gridtown: there is a pothole on 4th ave between 5th st and 6th st. How many routes to work can you take avoiding that unsightly (and dangerous) stretch of road? Explain.

(d) The pothole has been repaired (phew) and a new donut shop has opened on the corner of 4th ave and 5th st. How many routes to work drive by one or the other (or both) donut shops? Hint: the donut shops serve PIE.

12. **Suppose you are ordering a large pizza from D.P. Dough. You want 3 distinct toppings, chosen from their list of 11 vegetarian toppings.**

(a) How many choices do you have for your pizza?

(b) How many choices do you have for your pizza if you refuse to have pineapple as one of your toppings?

(c) How many choices do you have for your pizza if you insist on having pineapple as one of your toppings?

(d) How do the three questions above relate to each other? Explain.

Explain why the coefficient of $x^5 y^3$ the same as the coefficient of $x^3 y^5$ in the expansion of $(x + y)^8$?

2.3 COMBINATIONS AND PERMUTATIONS

Investigate!

You have a bunch of chips which come in five different colors: red, blue, green, purple and yellow.

- How many different two-chip stacks can you make if the bottom chip must be red or blue? Explain your answer using both the additive and multiplicative principles.
- How many different three-chip stacks can you make if the bottom chip must be red or blue and the top chip must be green, purple or yellow? How does this problem relate to the previous one?
- How many different three-chip stacks are there in which no color is repeated? What about four-chip stacks?
- Suppose you wanted to take three different colored chips and put them in your pocket. How many different choices do you have? What if you wanted four different colored chips? How do these problems relate to the previous one?

Attempt the above activity before proceeding

A permutation is a (possible) rearrangement of objects. For example, there are 6 permutations of the letters *a*, *b*, *c*:

$$abc, \quad acb, \quad bac, \quad bca, \quad cab, \quad cba.$$

We know that we have them all listed above —there are 3 choices for which letter we put first, then 2 choices for which letter comes next, which leaves only 1 choice for the last letter. The multiplicative principle says we multiply $3 \cdot 2 \cdot 1$.

Example 2.3.1

How many permutations are there of the letters a, b, c, d, e, f?

Solution. We do NOT want to try to list all of these out. However, if we did, we would need to pick a letter to write down first. There are 6 choices for that letter. For each choice of first letter, there are 5 choices for the second letter (we cannot repeat the first letter; we are rearranging letters and only have one of each), and for each of those, there are 4 choices for the third, 3 choices for the fourth, 2 choices for the fifth and finally only 1 choice for the last letter. So there are $6 \cdot 5 \cdot 4 \cdot 3 \cdot 2 \cdot 1 = 720$ permutations of the 6 letters.

A piece of notation is helpful here: $n!$, read "n factorial", is the product of all positive integers less than or equal to n (for reasons of convenience, we also define 0! to be 1). So the number of permutation of 6 letters, as seen in the previous example is $6! = 6 \cdot 5 \cdot 4 \cdot 3 \cdot 2 \cdot 1$. This generalizes:

Permutations of n elements

There are $n! = n \cdot (n - 1) \cdot (n - 2)\ 2 \cdot 1$ permutations of n (distinct) elements.

Example 1.3.2 Counting Bijective Functions

How many functions $f: \{1, 2, \ldots, 8\} \rightarrow \{1, 2, \ldots, 8\}$ are *bijective*?

Solution

Remember what it means for a function to be bijective: each element in the codomain must be the image of exactly one element of the domain. Using two-line notation,

$$f = \begin{pmatrix} 1 & 2 & 3 & 4 & 5 & 6 & 7 & 8 \\ 3 & 1 & 5 & 8 & 7 & 6 & 2 & 4 \end{pmatrix}.$$

What we are really doing is just rearranging the elements of the codomain, so we are creating a permutation of 8 elements. In fact, "permutation" is another term used to describe bijective functions from a finite set to itself.

If you believe this, then you see the answer must be $8! = 8 \cdot 7 \cdot \cdots \cdot 1 = 40320$. You can see this directly as well: for each element of the domain, we must pick a distinct element of the codomain to map to. There are 8 choices for where to send 1, then 7 choices for where to send 2, and so on. We multiply using the multiplicative principle.

Sometimes we do not want to permute all of the letters/numbers/ elements we are given.

Example 2.3.3

How many 4 letter "words" can you make from the letters a through *f*, with no repeated letters?

Solution

This is just like the problem of permuting 4 letters, only now we have more choices for each letter. For the first letter, there are 6 choices. For each of those, there are 5 choices for the second letter. Then there are 4 choices for the third letter, and 3 choices for the last letter. The total number of words is $6 \cdot 5 \cdot 4 \cdot 3 = 360$. This is not 6! because we never multiplied by 2 and 1. We could start with 6! and then cancel the 2 and 1, and thus write $\frac{6!}{2!}$.

In general, we can ask how many permutations exist of k objects choosing those objects from a larger collection of n objects. (In the example above, $k = 4$, and $n = 6$.) We write this number P n, k and sometimes call it a k-permutation of n elements. From the example above, we see that to compute $P(n, k)$ we must apply the multiplicative principle to k numbers, starting with n and counting backwards. For example

$$P(10, 4) = 10 \cdot 9 \cdot 8 \cdot 7.$$

Notice again that $P(10, 4)$ starts out looking like 10!, but we stop after 7. We can formally account for this "stopping" by dividing away the part of the factorial we do not want:

$$P(10, 4) = \frac{10 \cdot 9 \cdot 8 \cdot 7 \cdot 6 \cdot 5 \cdot 4 \cdot 3 \cdot 2 \cdot 1}{6 \cdot 5 \cdot 4 \cdot 3 \cdot 2 \cdot 1} = \frac{10!}{6!}.$$

Careful: The factorial in the denominator is not 4! but rather (10 - 4)!.

K-permutations **of *n* elements.**

$P(n, k)$ is the number of *k*-permutations of *n* elements, the number of ways to arrange k objects chosen from n distinct objects.

$$P(n, k) = \frac{n!}{(n-k)!} = n(n-1)(n-2)\cdots(n-(k-1)).$$

Note that when n = k, we have $P(n, n) = \frac{n!}{(n-n)!} = n!$ (since we defined 0! to be 1). This makes sense —we already know n! gives the number of permutations of all n objects.

Example 2.3.4 Counting injective functions

How many functions $f\colon \{1, 2, 3\} \to \{1, 2, 3, 4, 5, 6, 7, 8\}$ are injective?

Solution

Note that it doesn't make sense to ask for the number of bijections here, as there are none (because the codomain is larger than the domain, there are no surjections). But for a function to be injective, we just can't use an element of the codomain more than once.

We need to pick an element from the codomain to be the image of 1. There are 8 choices. Then we need to pick one of the remaining 7 elements to be the image of 2. Finally, one of the remaining 6 elements must be the image of 3. So the total number of functions is $8 \cdot 7 \cdot 6 = P(8, 3)$.

What this demonstrates in general is that the number of injections $f: A \to B$, where $|A| = k$ and $|B| = n$, is $P(n, k)$.

Here is another way to find the number of k-permutations of n elements: first select which k elements will be in the permutation, then count how many ways $\binom{n}{k}$ re are to arrange them. Once you have selected the k objects, we know there are ways to arrange (permute) them. But how do you select k objects from the n? You have n objects, and you need to choose k of them. You can do that in n ways. Then for each choice of those k elements, we can permute them in k! ways. Using the multiplicative principle, we get another formula for P(n, k):

$$P(n, k) = \binom{n}{k} \cdot k!.$$

Now since we have a closed formula for $P(n, k)$ already, we can substitute that in:

$$\frac{n!}{(n-k)!} = \binom{n}{k} \cdot k!.$$

If we divide both sides by ! we get a closed formula for $\binom{n}{k}$.

Closed formula for $\binom{n}{k}$.

$$\binom{n}{k} = \frac{n!}{(n-k)!k!} = \frac{n(n-1)(n-2)\cdots(n-(k-1))}{k(k-1)(k-2)\cdots 1}.$$

We say $P(n, k)$ counts permutations, and $\binom{n}{k}$ counts *combinations.* The formulas for each are very similar, there is just an extra k! in the denominator of $\binom{n}{k}$. That extra k! accounts for the fact that $\binom{n}{k}$ does not distinguish between the different orders that the k objects can appear in. We are just selecting (or choosing) the k objects, not arranging them. Perhaps "combination" is a misleading label. We don't mean it like a combination lock (where the order would definitely matter). Perhaps a better metaphor is a combination of flavors — you just need to decide which flavors to combine, not the order in which to combine them.

To further illustrate the connection between combinations and permutations, we close with an example.

Example 2.3.5

You decide to have a dinner party. Even though you are incredibly popular and have 14 different friends, you only have enough chairs to invite 6 of them.

- How many choices do you have for which 6 friends to invite?
- What if you need to decide not only which friends to invite but also where to seat them along your long table? How many choices do you have then?

Solution

You must simply choose 6 friends from a group of 14. This can be done in $\binom{14}{6}$ ways. We can find this number either by using Pascal's triangle or the closed formula: $\frac{14!}{8!\cdot 6!} = 3003$.

Here you must count all the ways you can permute 6 friends chosen from a group of 14. So the answer is $P(14, 6)$, which can be calculated as $\frac{14!}{8!} = 2162160$.

Notice that we can think of this counting problem as a question about counting functions: how many injective functions are there from your set of 6 chairs to your set of 14 friends (the functions are injective because you can't have a single chair go to two of your friends).

How are these numbers related? Notice that $P(14,6)$ is *much* larger than $\binom{14}{6}$. This makes sense. $\binom{14}{6}$. picks 6 friends, but $P(14,6)$ arranges the 6 friends as well as picks them. In fact, we can say exactly how much larger $P(14, 6)$ is. In both counting problems we choose 6 out of 14 friends. For the first one, we stop there, at 3003 ways. But for the second counting problem, each of those 3003 choices of 6 friends can be arranged in exactly 6! ways. So now we have 3003.6! choices and that is exactly 2162160.

Alternatively, look at the first problem another way. We want to select 6 out of 14 friends, but we do not care about the order they are selected in. To select 6 out of 14 friends, we might try this: 14.13.12.11.10.9.

This is a reasonable guess, since we have 14 choices for the first guest, then 13 for the second, and so on. But the guess is wrong (in fact, that product is exactly $2162160 = P(14, 6)$). It distinguishes between the different orders in which we could invite the guests.

To correct for this, we could divide by the number of different arrangements of the 6 guests (so that all of these would count as just one outcome). There are precisely 6! ways to arrange 6 guests, so the correct answer to the first question is

$$\frac{14\cdot 13\cdot 12\cdot 11\cdot 10\cdot 9}{6!}.$$

Note that another way to write this is which is what we had originally.

$$\frac{14!}{8! \cdot 6!}.$$

EXERCISES

1. **A pizza parlor offers 10 toppings.**

   (a) How many 3-topping pizzas could they put on their menu? Assume double toppings are not allowed.

   (b) How many total pizzas are possible, with between zero and ten toppings (but not double toppings) allowed?

   (c) The pizza parlor will list the 10 toppings in two equal-sized columns on their menu. How many ways can they arrange the toppings in the left column?

2. **A combination lock consists of a dial with 40 numbers on it. To open the lock, you turn the dial to the right until you reach a first number, then to the left until you get to second number, then to the right again to the third number. The numbers must be distinct. How many different combinations are possible?**

3. **Using the digits 2 through 8, find the number of different 5-digit numbers such that:**

   (a) Digits can be used more than once.

   (b) Digits cannot be repeated, but can come in any order.

   (c) Digits cannot be repeated and must be written in increasing order.

   (d) Which of the above counting questions is a combination and which is a permutation? Explain why this makes sense.

4. **In an attempt to clean up your room, you have purchased a new floating shelf to put some of your 17 books you have stacked in a corner. These books are all by different authors. The new book shelf is large enough to hold 10 of the books.**

   (a) How many ways can you select and arrange 10 of the 17 books on the shelf? Notice that here we will allow the books to end up in any order. Explain.

   (b) How many ways can you arrange 10 of the 17 books on the shelf if you insist they must be arranged alphabetically by author? Explain.

5. **Suppose you wanted to draw a quadrilateral using the dots below as vertices (corners). The dots are spaced one unit apart horizontally and two units apart vertically.**

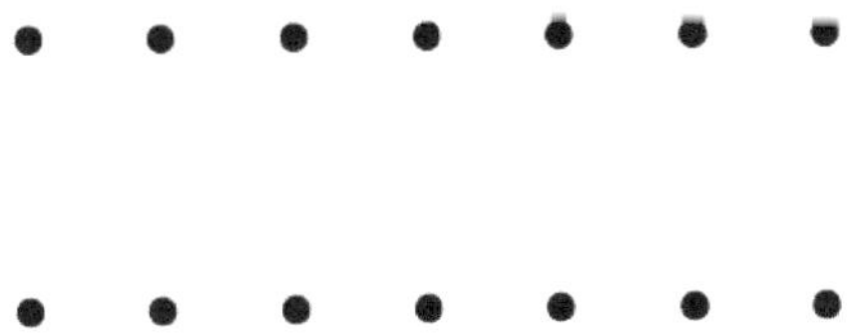

How many quadrilaterals are possible?

How many are squares?

How many are rectangles?

How many are parallelograms?

How many are trapezoids? (Here, as in calculus, a trapezoid is defined as a quadrilateral with *at least* one pair of parallel sides. In particular, parallelograms are trapezoids.)

How many are trapezoids that are not parallelograms?

6. **How many triangles are there with vertices from the points shown below? Note, we are not allowing degenerate triangles-ones with all three vertices on the same line, but we do allow non-right triangles. Explain why your answer is correct.**

7. **An *anagram* of a word is just a rearrangement of its letters. How many different anagrams of "uncopyrightable" are there? (This happens to be the longest common English word without any repeated letters.)**
8. **How many anagrams are there of the word "assesses" that start with the letter "a"?**
9. **How many anagrams are there of "anagram"?**
10. **On a business retreat, your company of 20 executives go golfing.**
    (a) You need to divide up into foursomes (groups of 4 people): a first foursome, a second foursome, and so on. How many ways can you do this?
    (b) After all your hard work, you realize that in fact, you want each foursome to include one of the five Board members. How many ways can you do this?

11. **How many different seating arrangements are possible for King Arthur and his 9 knights around their round table?**
12. **Consider sets A and B with $|A| = 10$ and $|B| = 17$.**
    (a) How many functions $f: A \to B$ are there?
    (b) How many functions $f: A \to B$ are injective?
13. **Consider functions $f: \{1, 2, 3, 4\} \to \{1, 2, 3, 4, 5, 6\}$.**
    (a) How many functions are there total?
    (b) How many functions are injective?
    (c) How many of the injective functions are increasing? To be increasing means that if $a < b$ then $f(a) < f(b)$, or in other words, the outputs get larger as the inputs get larger.
14. **We have seen that the formula for** $P(n, k)$ is $\frac{n!}{(n-k)!}$. **Your task here is to explain why this is the right formula.**
    (a) Suppose you have 12 chips, each a different color. How many different stacks of 5 chips can you make? Explain your answer and why it is the same as using the formula for $P(12, 5)$.
    (b) Using the scenario of the 12 chips again, what does 12! count? What does 7! count? Explain.
    (c) Explain why it makes sense to divide 12! by 7! when computing P(12, 5) (in terms of the chips).
    (d) Does your explanation work for numbers other than 12 and 5? Explain the formula $P(n, k) = \frac{n!}{(n-k)!}$.using the variables n and k.

2.4 COMBINATORIAL PROOFS

Investigate!

1. **The Stanley Cup is decided in a best of 7 tournament between two teams. In how many ways can your team win? Let's answer this question two ways:**
    (a) How many of the 7 games does your team need to win? How many ways can this happen?
    (b) What if the tournament goes all 7 games? So you win the last game. How many ways can the first 6 games go down?
    (c) What if the tournament goes just 6 games? How many ways can this happen? What about 5 games? 4 games?
    (d) What are the two different ways to compute the number of ways your team can win? Write down an equation involving binomial coefficients (that is, $\binom{n}{k}$'s). What pattern in Pascal's triangle is this an example of?

2. **Generalize. What if the rules changed and you played a best of 9 tournament (5 wins required)? What if you played an n game tournament with k wins required to be named champion?**

Attempt the above activity before proceeding

Patterns in Pascal's Triangle

Have a look again at Pascal's triangle. Forget for a moment where it comes from. Just look at it as a mathematical object. What do you notice?

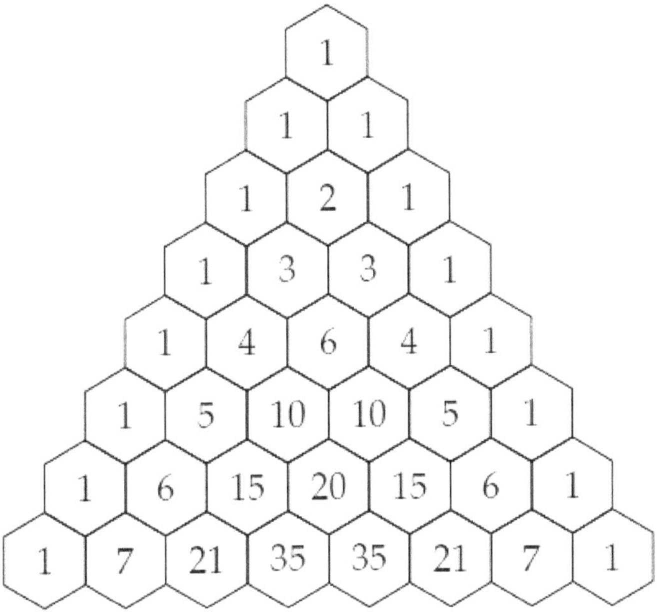

There are lots of patterns hidden away in the triangle, enough to fill a reasonably sized book. Here are just a few of the most obvious ones:

- The entries on the border of the triangle are all 1.
- Any entry not on the border is the sum of the two entries above it.
- The triangle is symmetric. In any row, entries on the left side are mirrored on the right side.
- The sum of all entries on a given row is a power of 2. (You should check this!)

We would like to state these observations in a more precise way, and then prove that they are correct. Now each entry in Pascal's triangle is in fact a binomial coefficient. The 1 on the very top of the triangle is $\binom{0}{0}$. The next row (which we will call row 1, even though it is not the top-most row) consists of $\binom{1}{0}$ and $\binom{1}{1}$. Row 4 (the row 1, 4, 6, 4, 1) consists of the binomial coefficients

$$\binom{4}{0}\ \binom{4}{1}\ \binom{4}{2}\ \binom{4}{3}\ \binom{4}{4}.$$

Given this description of the elements in Pascal's triangle, we can rewrite the above observations as follows:

- $\binom{n}{0} = 1$ and $\binom{n}{n} = 1$.
- $\binom{n}{k} = \binom{n-1}{k-1} + \binom{n-1}{k}$.
- $\binom{n}{k} = \binom{n}{n-k}$.
- $\binom{n}{0} + \binom{n}{1} + \binom{n}{2} + \cdots + \binom{n}{n} = 2^n$.

Each of these is an example of a **binomial identity** : an identity (i.e., equation) involving binomial coefficients.

Our goal is to establish these identities. We wish to prove that they hold for all values of *n* and *k*. These proofs can be done in many ways. One option would be to give algebraic proofs, using the formula for $\binom{n}{k}$:

$$\binom{n}{k} = \frac{n!}{(n-k)!\,k!}.$$

Here's how you might do that for the second identity above.

Example 2.4.1

Give an algebraic proof for the binomial identity

$$\binom{n}{k} = \binom{n-1}{k-1} + \binom{n-1}{k}.$$

Solution.

Proof. By the definition of $\binom{n}{k}$, we have

$$\binom{n-1}{k-1} = \frac{(n-1)!}{(n-1-(k-1))!(k-1)!} = \frac{(n-1)!}{(n-k)!(k-1)!}$$

and

$$\binom{n-1}{k} = \frac{(n-1)!}{(n-1-k)!k!}.$$

Thus, starting with the right-hand side of the equation:

$$\begin{aligned}\binom{n-1}{k-1}+\binom{n-1}{k} &= \frac{(n-1)!}{(n-k)!(k-1)!}+\frac{(n-1)!}{(n-1-k)!\,k!}\\ &= \frac{(n-1)!k}{(n-k)!\,k!}+\frac{(n-1)!(n-k)}{(n-k)!\,k!}\\ &= \frac{(n-1)!(k+n-k)}{(n-k)!\,k!}\\ &= \frac{n!}{(n-k)!\,k!}\\ &= \binom{n}{k}.\end{aligned}$$

The second line (where the common denominator is found) works because $k(k - 1)! = k!$ and $(n - k)(n - k - 1)! = (n - k)!$.

This is certainly a valid proof, but also is entirely useless. Even if you understand the proof perfectly, it does not tell you why the identity is true. A better approach would be to explain what $\binom{n}{k}$ means and then say why that is also what $\binom{n-1}{k-1}+\binom{n-1}{k}$ means. Let's see how this works for the four identities we observed above.

Example 2.4.2

Explain why $\binom{n}{0} = 1$.and $\binom{n}{n} = 1$.

Solution.

What do these binomial coefficients tell us? Well, $\binom{n}{0}$ gives the number of ways to select 0 objects from a collection of n objects. There is only one way to do this, namely to not select any of the objects. Thus $\binom{n}{0} = 1$. Similarly, $\binom{n}{n}$ gives the number of ways to select n objects from a collection of n objects. There is only one way to do this: select all n objects. Thus $\binom{n}{n} = 1$.

Alternatively, we know that $\binom{n}{0}$ is the number of n-bit strings with weight 0. There is only one such string, the string of all 0's. So $\binom{n}{0} = 1$. Similarly is the number of n-bit strings with weight n. There is only one string with this property, the string of all 1's.

Another way: $\binom{n}{0}$ gives the number of subsets of a set of size n containing 0 elements. There is only one such subset, the empty set. $\binom{n}{n}$ gives the number of subsets containing n elements. The only such subset is the original set (of all elements).

Example 2.4.3

Explain

$$\binom{n}{k} = \binom{n-1}{k-1} + \binom{n-1}{k}.$$

Solution. The easiest way to see this is to consider bit strings. $\binom{n}{k}$ is the number of bit strings of length n containing k 1's. Of all of these strings, some start with a 1 and the rest start with a 0. First consider all the bit strings which start with a 1. After the 1, there must be n - 1 more bits (to get the total length up to n) and exactly k - 1 of them must be 1's (as we already have one, and we need k total). How many strings are there like that? There are exactly $\binom{n-1}{k-1}$ such bit strings, so of all the length n bit strings containing k 1's, $\binom{n-1}{k-1}$ of them start with a 1. Similarly, there are $\binom{n-1}{k}$ which start with a 0 (we still need n - 1 bits and now k of them must be 1's).

Since there are $\binom{n-1}{k}$ bit strings containing n^{-1} bits with k 1's, that is the number of length n bit strings with k 1's which start with a 0.

Therefore $\binom{n}{k} = \binom{n-1}{k-1} + \binom{n-1}{k}$.

Another way: consider the question, how many ways can you select k pizza toppings from a menu containing n choices? One way to do this is just $\binom{n}{k}$.Another way to answer the same question is to first decide whether or not you want anchovies. If you do want anchovies, you still need to pick k - 1 toppings, now from just n - 1 choices. That can be done in $\binom{n-1}{k-1}$ ways. If you do not want anchovies, then you still need to select k toppings from n - 1 choices (the anchovies are out). You can do that in $\binom{n-1}{k}$ ways. Since the choices with anchovies are disjoint from the choices without anchovies, the total choices are $\binom{n-1}{k-1} + \binom{n-1}{k}$.But wait. We answered the same question in two different ways, so the two answers must be the same. Thus

$$\binom{n}{k} = \binom{n-1}{k-1} + \binom{n-1}{k}.$$

You can also explain (prove) this identity by counting subsets, or even lattice paths.

Example 2.4.4

Prove the binomial identity $\binom{n}{k} = \binom{n}{n-k}$.

Solution. Why is this true? $\binom{n}{k}$ counts the number of ways to select k things from n choices. On the other hand $\binom{n}{n-k}$ counts the number of ways to select n k things from n choices. Are these really the same? Well, what if instead of selecting the n - k things you choose to exclude them. How many ways are there to choose things to exclude from choices. Clearly this is $\binom{n}{n-k}$ as well (it doesn't matter whether you include or exclude the things once you have chosen them). And if you exclude n k things, then you are including the other k things. So the set of outcomes should be the same.

Let's try the pizza counting example like we did above. How many ways are there to pick k toppings from a list of n choices? On the one hand, the answer is simply $\binom{n}{k}$. Alternatively, you could make a list of all the toppings you don't want. To end up with a pizza containing exactly k toppings, you need to pick n - k toppings to not put on the pizza. You have $\binom{n}{n-k}$ choices for the toppings you don't want. Both of these ways give you a pizza with k toppings, in fact all the ways to get a pizza with k toppings. Thus these two answers must be the same: $\binom{n}{k} = \binom{n}{n-k}$.

You can also prove (explain) this identity using bit strings, subsets, or lattice paths. The bit string argument is nice: $\binom{n}{k}$ counts the number of bit strings of length n with k 1's. This is also the number of bit string of length n with k 0's (just replace each 1 with a 0 and each 0 with a 1). But if a string of length n has k 0's, it must have n - k 1's. And there are exactly $\binom{n}{n-k}$ strings of length n with n – k 1's.

Example 2.4.5

Prove the binomial identity $\binom{n}{0} + \binom{n}{1} + \binom{n}{2} + \cdots + \binom{n}{n} = 2^n$.

Solution

Let's do a "pizza proof" again. We need to find a question about pizza toppings which has 2^n as the answer. How about this: If a pizza joint offers n toppings, how many pizzas can you build using any number of toppings from no toppings to all toppings, using each topping at most once?

On one hand, the answer is 2^n . For each topping you can say "yes" or "no," so you have two choices for each topping.

On the other hand, divide the possible pizzas into disjoint groups: the pizzas with no toppings, the pizzas with one topping, the pizzas with two toppings, etc. If we want no toppings, there is only one pizza like that (the empty pizza, if you will) but it would be better to think of that number as $\binom{n}{0}$ since we choose 0 of the n toppings. How many pizzas have 1 topping? We need to choose 1 of the n toppings, so $\binom{n}{1}$. We have:

- Pizzas with 0 toppings: $\binom{n}{0}$
- Pizzas with 1 topping: $\binom{n}{1}$
- Pizzas with 2 toppings: $\binom{n}{2}$
- $\vdots$
- Pizzas with n toppings: $\binom{n}{n}$.

The total number of possible pizzas will be the sum of these, which is exactly the left-hand side of the identity we are trying to prove.

Again, we could have proved the identity using subsets, bit strings, or lattice paths (although the lattice path argument is a little tricky).

Hopefully this gives some idea of how explanatory proofs of binomial identities can go. It is worth pointing out that more traditional proofs can also be beautiful. For example, consider the following rather slick proof of the last identity.

Expand the binomial $(x + y)^n$:

$$(x+y)^n = \binom{n}{0}x^n + \binom{n}{1}x^{n-1}y + \binom{n}{2}x^{n-2}y^2 + \cdots + \binom{n}{n-1}x \cdot y^n + \binom{n}{n}y^n.$$

Let $x = 1$ and $y = 1$. We get:

$$(1+1)^n = \binom{n}{0}1^n + \binom{n}{1}1^{n-1}1 + \binom{n}{2}1^{n-2}1^2 + \cdots + \binom{n}{n-1}1 \cdot 1^n + \binom{n}{n}1^n.$$

Of course this simplifies to:

$$(2)^n = \binom{n}{0} + \binom{n}{1} + \binom{n}{2} + \cdots + \binom{n}{n-1} + \binom{n}{n}.$$

Something fun to try: Let x = 1 and y = 2. Neat huh?

More Proofs

The explanatory proofs given in the above examples are typically called combinatorial proofs. In general, to give a combinatorial proof for a binomial identity, say $A = B$ you do the following:

- Find a counting problem you will be able to answer in two ways.
- Explain why one answer to the counting problem is A.
- Explain why the other answer to the counting problem is B.

Since both A and B are the answers to the same question, we must have A = B.

The tricky thing is coming up with the question. This is not always obvious, but it gets easier the more counting problems you solve. You will start to recognize types of answers as the answers to types of questions. More often what will happen is you will be solving a counting problem and happen to think up two different ways of finding the answer. Now you have a binomial identity and the proof is right there. The proof is the problem you just solved together with your two solutions.

For example, consider this counting question:

- How many 10-letter words use exactly four A's, three B's, two C's and one D?

Let's try to solve this problem. We have 10 spots for letters to go. Four of those need to be A's. We can pick the four A-spots in $\binom{10}{4}$ ways. Now where can we put the B's? Well there are only 6 spots left, we need to pick 3 of them. This can be done in $\binom{6}{3}$ ways. The two C's need to go in two of the 3 remaining spots, we have $\binom{3}{2}$ ways of doing that. That leaves just one spot of the D, but we could write that 1 choice as is: $\binom{1}{1}$. Thus the answer is:

$$\binom{10}{4}\binom{6}{3}\binom{3}{2}\binom{1}{1}.$$

But why stop there? We can find the answer another way too. First let's decide where to put the one D: we have 10 spots, we need to choose 1 of them, so this can be done in $\binom{10}{1}$ ways. Next, choose one of the $\binom{9}{2}$ ways to place the two C's. We now have 7 spots left, and three of them need to be filled with B's. There are $\binom{7}{3}$ ways to do this. Finally the A's can be placed in $\binom{4}{4}$ (that is, only one) ways. So another answer to the question is

$$\binom{10}{1}\binom{9}{2}\binom{7}{3}\binom{4}{4}.$$

Interesting. This gives us the binomial identity:

$$\binom{10}{4}\binom{6}{3}\binom{3}{2}\binom{1}{1} = \binom{10}{1}\binom{9}{2}\binom{7}{3}\binom{4}{4}.$$

Here are a couple more binomial identities with combinatorial proofs.

Example 2.4.6

Prove the identity

$$1n + 2(n-1) + 3(n-2) + \cdots + (n-1)2 + n1 = \binom{n+2}{3}.$$

Solution

To give a combinatorial proof we need to think up a question we can answer in two ways: one way needs to give the left-hand-side of the identity, the other way needs to be the right- hand-side of the identity. Our clue to what question to ask comes from the right-hand side: $\binom{n+2}{3}$ counts the number of ways to select 3 things from a group of n + 2 things. Let's name those things 1, 2, 3, . . . , $n + 2$. In other words, we want to find 3-element subsets of those numbers (since order should not matter, subsets are exactly the right thing to think about). We will have to be a bit clever to explain why the left-hand-side also gives the number of these subsets. Here's the proof.

Proof. Consider the question "How many 3-element subsets are there of the set $\{1, 2, 3, \ldots, n + 2\}$?" We answer this in two ways:

Answer 1: We must select 3 elements from the collection of n + 2 elements. This can be done in $\binom{n+2}{3}$ ways.

Answer 2: Break this problem up into cases by what the middle number in the subset is. Say each subset is $\{a, b, c\}$ written in increasing order. We count the number of subsets for each distinct value of b. The smallest possible value of b is 2, and the largest is $n + 1$.

When b = 2, there are 1 n subsets: 1 choice for a and n choices (3 through $n + 2$) for *c*.

When $b = 3$, there are 2 (n- 1) subsets: 2 choices for a and n- 1 choices for *c*.

When $b = 4$, there are 3 (n- 2) subsets: 3 choices for a and n 2 choices for *c*.

And so on. When b = n + 1, there are n choices for a and only 1 choice for *c*, so n 1 subsets.

Therefore the total number of subsets is

$$1n + 2(n - 1) + 3(n - 2) + \cdots + (n - 1)2 + n1.$$

Since Answer 1 and Answer 2 are answers to the same question, they must be equal. Therefore

$$1n + 2(n - 1) + 3(n - 2) + \cdots + (n - 1)2 + n1 = \binom{n+2}{3}.$$

Example 2.4.7

Prove the binomial identity

$$\binom{n}{0}^2 + \binom{n}{1}^2 + \binom{n}{2}^2 + \cdots + \binom{n}{n}^2 = \binom{2n}{n}.$$

Solution

We will give two different proofs of this fact. The first will be very similar to the previous example (counting subsets). The second proof is a little slicker, using lattice paths.

Proof. **Consider the question:** "How many pizzas can you make using n toppings when there are 2n toppings to choose from?"

Answer 1: There are 2n toppings, from which you must choose *n*. This can be done in $\binom{2n}{n}$ ways.

Answer 2: Divide the toppings into two groups of *n* toppings (perhaps n meats and n veggies). Any choice of n toppings must include some number from the first group and some number from the second group. Consider each possible number of meat toppings separately:

- **0 meats:** $\binom{n}{0}\binom{n}{n}$, since you need to choose 0 of the n meats and n veggies.
- **1 meats:** $\binom{n}{1}\binom{n}{n-1}$, since you need 1 of n meats so n-1 of n veggies.
- **2 meats:** $\binom{n}{2}\binom{n}{n-2}$. Choose 2 meats and the remaining n-2 toppings from the n veggies.

And so on. The last case is n meats, which can be done in $\binom{n}{n}\binom{n}{0}$ ways.

Thus the total number of pizzas possible is

$$\binom{n}{0}\binom{n}{n}+\binom{n}{1}\binom{n}{n-1}+\binom{n}{2}\binom{n}{n-2}+\cdots+\binom{n}{n}\binom{n}{0}.$$

This is not quite the left-hand side . . . yet. Notice that $\binom{n}{n}=\binom{n}{0}$ and $\binom{n}{n-1}=\binom{n}{1}$ and so on, by the identify in Example 1.4.4. Thus

we do indeed get

$$\binom{n}{0}^2+\binom{n}{1}^2+\binom{n}{2}^2+\cdots+\binom{n}{n}^2.$$

Since these two answers are answers to the same question, they must be equal, and thus

$$\binom{n}{0}^2+\binom{n}{1}^2+\binom{n}{2}^2+\cdots+\binom{n}{n}^2=\binom{2n}{n}.$$

For an alternative proof, we use lattice paths. This is reasonable to consider because the right-hand side of the identity reminds us of the number of paths from (0, 0) to (n, n).

***Proof.* Consider the question:** How many lattice paths are there from (0, 0) to (n, n) ?

Answer 1: We must travel 2n steps, and n of them must be in the up direction. Thus there are $\binom{2n}{n}$ paths.

Answer 2: Note that any path from (0, 0) to (n, n) must cross the line x + y = n. That is, any path must pass through exactly one of the points: $(0, n)$, $(1, n-1)$, $(2, n-2)$, . . . , $(n, 0)$. For example, this is what happens in the case n = 4:

How many paths pass through (0, n)? To get to that point, you must travel n units, and 0 of them are to the right, so there are $\binom{n}{0}$ ways to get to (0, n). From (0, n) to (n, n) takes n steps, and 0 of them are up. So there are $\binom{n}{0}$ ways to get from (0, n) to (n, n). Therefore there are $\binom{n}{0}\binom{n}{0}$ paths from ¹0; 0º to ¹n; nº through the point (0, n).

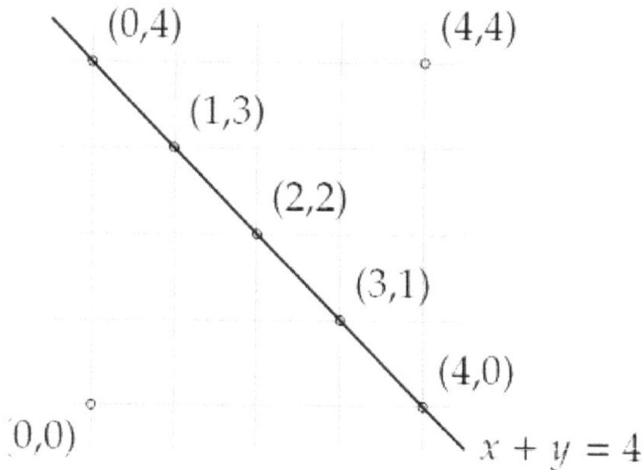

What about through (1,n-1). There are $\binom{n}{1}$ paths to get there (n steps, 1 to the right) and $\binom{n}{1}$ paths to complete the journey to (n, n) (n steps, 1 up). So there are $\binom{n}{1}\binom{n}{1}$ paths from (0, 0) to (n, n) through (1, n^{-1}).

What about through (1, n^{-1}). There are 1 paths to get there (*n*, *n*) (n steps, 1 up). So there are through (1, n^{-1}).

In general, to get to (n, n) through the point (*k*, *n* - *k*) we have $\binom{n}{k}$ paths from the midpoint and then $\binom{n}{k}$ paths from the midpoint to (*n*, *n*). So there are $\binom{n}{k}\binom{n}{k}$ paths from (0, 0) to (n, n) through (*k*, *n* - *k*).

All together then the total paths from (0, 0) to (*n*, *n*) passing through exactly one of these midpoints is

$$\binom{n}{0}^2 + \binom{n}{1}^2 + \binom{n}{2}^2 + \cdots + \binom{n}{n}^2.$$

Since these two answers are answers to the same question, they must be equal, and thus

$$\binom{n}{0}^2 + \binom{n}{1}^2 + \binom{n}{2}^2 + \cdots + \binom{n}{n}^2 = \binom{2n}{n}.$$

EXERCISES

1. **Give a combinatorial proof of the identity 2 + 2 + 2 = 3 · 2.**
2. **Suppose you own x fezzes and y bow ties. Of course, x and y are both greater than 1.**
   (a) How many combinations of fez and bow tie can you make? You can wear only one fez and one bow tie at a time. Explain.

   (b) Explain why the answer is also $\binom{x+y}{2} - \binom{x}{2} - \binom{y}{2}$. (If this is what you claimed the answer was in part (a), try it again.)

(c) Use your answers to parts (a) and (b) to give a combinatorial proof of the identity

$$\binom{x+y}{2} - \binom{x}{2} - \binom{y}{2} = xy..$$

3. How many triangles can you draw using the dots below as vertices?

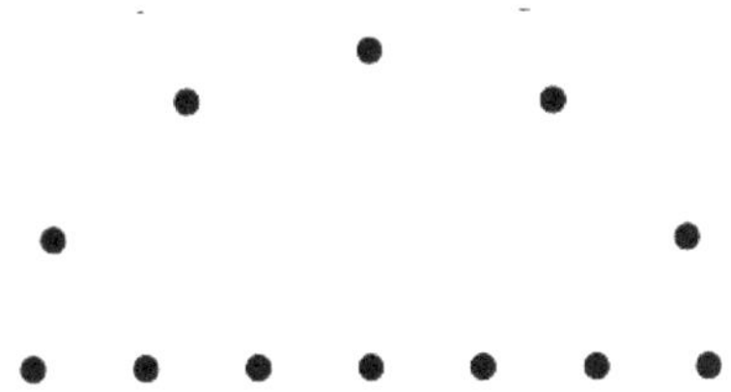

(a) Find an expression for the answer which is the sum of three terms involving binomial coefficients.

(b) Find an expression for the answer which is the difference of two binomial coefficients.

(c) Generalize the above to state and prove a binomial identity using a combinatorial proof. Say you have x points on the horizontal axis and y points in the semi-circle.

4. Consider all the triangles you can create using the points shown below as vertices. Note, we are not allowing degenerate triangles (ones with all three vertices on the same line) but we do allow non-right triangles.

(a) Find the number of triangles, and explain why your answer is correct.

(b) Find the number of triangles again, using a different method. Explain why your new method works.

(c) State a binomial identity that your two answers above establish (that is, give the binomial identity that your two answers a proof for). Then generalize this using *m*'s and *n*'s.

5. A woman is getting married. She has 15 best friends but can only select 6 of them to be her bridesmaids, one of which needs to be her maid of honor. How many ways can she do this?

(a) What if she first selects the 6 bridesmaids, and then selects one of them to be the maid of honor?

(b) What if she first selects her maid of honor, and then 5 other bridesmaids?

(c) Explain why $6\binom{15}{6} = 15\binom{14}{5}$.

6. **Consider the identity:**

$$k\binom{n}{k} = n\binom{n-1}{k-1}.$$

(a) Is this true? Try it for a few values of *n* and *k*.

(b) Use the formula for $\binom{n}{k}$ to give an algebraic proof of the identity.

(c) Give a combinatorial proof of the identity.

7. **Give a combinatorial proof of the identity** $\binom{n}{2}\binom{n-2}{k-2} = \binom{n}{k}\binom{k}{2}$.

8. **Consider the binomial identity**

$$\binom{n}{1} + 2\binom{n}{2} + 3\binom{n}{3} + \cdots + n\binom{n}{n} = n2^{n-1}.$$

(a) Give a combinatorial proof of this identity. Hint: What if some number of a group of n people wanted to go to an escape room, and among those going, one needed to be the team captain?

(b) Give an alternate proof by multiplying out $(1 + x)^n$ and taking derivatives of both sides.

9. **Give a combinatorial proof for the identity** $1 + 2 + 3 + \cdots + n = \binom{n+1}{2}$.

10. **Consider the bit strings in $\mathbf{B}^6_2$ (bit strings of length 6 and weight 2).**

(a) How many of those bit strings start with 1?

(b) How many of those bit strings start with 01?

(c) How many of those bit strings start with 001?

(d) Are there any other strings we have not counted yet? Which ones, and how many are there?

(e) How many bit strings are there total in $\mathbf{B}^6_2$?

(f) What binomial identity have you just given a combinatorial proof for?.

11. **Let's count ternary digit strings, that is, strings in which each digit can be 0,1, or 2.**

(a) How many ternary digit strings contain exactly *n* digits?

(b) How many ternary digit strings contain exactly *n* digits and n 2's.

(c) How many ternary digit strings contain exactly *n* digits and n- 1 2's. (Hint: where can you put the non-2 digit, and then what could it be?)

(d) How many ternary digit strings contain exactly n digits and n- 2 2's. (Hint: see previous hint)

(e) How many ternary digit strings contain exactly n digits and n - k 2's.

(f) How many ternary digit strings contain exactly n digits and no 2's. (Hint: what kind of a string is this?)

(g) Use the above parts to give a combinatorial proof for the identity

$$\binom{n}{0}+2\binom{n}{1}+2^2\binom{n}{2}+2^3\binom{n}{3}+\cdots+2^n\binom{n}{n}=3^n.$$

12. How many ways are there to rearrange the letters in the word "rear- range"? Answer this question in at least two different ways to establish a binomial identity.

13. Establish the identity below using a combinatorial proof.

$$\binom{2}{2}\binom{n}{2}+\binom{3}{2}\binom{n-1}{2}+\binom{4}{2}\binom{n-2}{2}+\cdots+\binom{n}{2}\binom{2}{2}=\binom{n+3}{5}.$$

14. In Example 1.4.5 we established that the sum of any row in Pascal's triangle is a power of two. Specifically,

$$\binom{n}{0}+\binom{n}{1}+\binom{n}{2}+\cdots+\binom{n}{n}=2^n.$$

The argument given there used the counting question, "how many pizzas can you build using any number of n different toppings?" To practice, give new proofs of this identity using different questions.

(a) Use a question about counting subsets.

(b) Use a question about counting bit strings.

(c) Use a question about counting lattice paths.

2.5 STARS AND BARS

Investigate!

Suppose you have some number of identical Rubik's cubes to distribute to your friends. Imagine you start with a single row of the cubes.

1. Find the number of different ways you can distribute the cubes provided:

(a) You have 3 cubes to give to 2 people.

(b) You have 4 cubes to give to 2 people.

(c) You have 5 cubes to give to 2 people.

(d) You have 3 cubes to give to 3 people.

(e) You have 4 cubes to give to 3 people.

(f) You have 5 cubes to give to 3 people.

2. **Make a conjecture about how many different ways you could distribute 7 cubes to 4 people. Explain.**

3. **What if each person were required to get** *at least* **one cube? How would your answers change?**

Attempt the above activity before proceeding

Consider the following counting problem:

- You have 7 cookies to give to 4 kids. How many ways can you do this?

Take a moment to think about how you might solve this problem. You may assume that it is acceptable to give a kid no cookies. Also, the cookies are all identical and the order in which you give out the cookies does not matter.

Before solving the problem, here is a wrong answer: You might guess that the answer should be 47 because for each of the 7 cookies, there are 4 choices of kids to which you can give the cookie. This is reasonable, but wrong. To see why, consider a few possible outcomes: we could assign the first six cookies to kid A, and the seventh cookie to kid B. Another outcome would assign the first cookie to kid B and the six remaining cookies to kid A. Both outcomes are included in the 4^7 answer. But for our counting problem, both outcomes are really the same – kid A gets six cookies and kid B gets one cookie.

What do outcomes actually look like? How can we represent them? One approach would be to write an outcome as a string of four numbers like this:

$$3112,$$

which represent the outcome in which the first kid gets 3 cookies, the second and third kid each get 1 cookie, and the fourth kid gets 2 cookies. Represented this way, the order in which the numbers occur matters. 1312 is a different outcome, because the first kid gets a one cookie instead of Each number in the string can be any integer between 0 and 7. But the answer is not 74. We need the *sum* of the numbers to be 7.

Another way we might represent outcomes is to write a string of seven letters:

ABAADCD,

which represents that the first cookie goes to kid A, the second cookie goes to kid B, the third and fourth cookies go to kid A, and so on. In fact, this outcome is identical to the previous one—A gets 3 cookies, B and C get 1 each and D gets 2. Each of the seven letters in the string can be any of the 4 possible letters (one for each kid), but the number of such strings is not 4^7, because here order does *not* matter. In fact, another way to write the same outcome is

AAABCDD.

This will be the preferred representation of the outcome. Since we can write the letters in any order, we might as well write them in alphabetical order for the purposes of counting. So we will write all the A's first, then all the B's, and so on.

Now think about how you could specify such an outcome. All we really need to do is say when to switch from one letter to the next. In terms of cookies, we need to say after how many cookies do we stop giving cookies to the first kid and start giving cookies to the second kid. And then after how many do we switch to the third kid? And after how many do we switch to the fourth? So yet another way to represent an outcome is like this:

$$***|*|*|**.$$

Three cookies go to the first kid, then we switch and give one cookie to the second kid, then switch, one to the third kid, switch, two to the fourth kid. Notice that we need 7 stars and 3 bars – one star for each cookie, and one bar for each switch between kids, so one fewer bars than there are kids (we don't need to switch after the last kid – we are done).

Why have we done all of this? Simple: to count the number of ways to distribute 7 cookies to 4 kids, all we need to do is count how many *stars and bars* charts there are. But a **stars and bars chart** is just a string of symbols, some stars and some bars. If instead of stars and bars we would use 0's and 1's, it would just be a bit string. We know how to count those.

Before we get too excited, we should make sure that really *any* string of (in our case) 7 stars and 3 bars corresponds to a different way to distribute cookies to kids. In particular consider a string like this:

$$|***||****.$$

Does that correspond to a cookie distribution? Yes. It represents the distribution in which kid A gets 0 cookies (because we switch to kid B before any stars), kid B gets three cookies (three stars before the next bar), kid C gets 0 cookies (no stars before the next bar) and kid D gets the remaining 4 cookies. No matter how the stars and bars are arranged, we can distribute cookies in that way. Also, given any way to distribute cookies, we can represent that with a stars and bars chart. For example, the distribution in which kid A gets 6 cookies and kid B gets 1 cookie has the following chart:

$$******|*||.$$

After all that work we are finally ready to count. Each way to distribute cookies corresponds to a stars and bars chart with 7 stars and 3 bars. So there are 10 symbols, and we must choose 3 of them to be bars. Thus:

There are $\binom{10}{3}$ ways to distribute 7 cookies to 4 kids.

While we are at it, we can also answer a related question: how many ways are there to distribute 7 cookies to 4 kids so that each kid gets at least one cookie? What can you say about the corresponding stars and bars charts? The charts must start and end with at least one star (so that kids A and D) get cookies, and also no two bars can be adjacent (so that kids B and C are not skipped). One way to assure this is to place bars only in the spaces between the stars. With 7 stars, there are 6 spots between the stars, so we must choose 3 of those 6 spots to fill with bars. Thus there are $\binom{6}{3}$ ways to distribute 7 cookies to 4 kids giving at least one cookie to each kid.

Another (and more general) way to approach this modified problem is to first give each kid one cookie. Now the remaining 3 cookies can be distributed to the 4 kids without restrictions. So we have 3 stars and 3 bars for a total of 6 symbols, 3 of which must be bars. So again we see that there are $\binom{6}{3}$ ways to distribute the cookies.

Stars and bars can be used in counting problems other than kids and cookies. Here are a few examples:

Example 2.5.1

Your favorite mathematical ice-cream parlor offers 10 flavors. How many milkshakes could you create using exactly 6, not necessarily distinct scoops? The order you add the flavors does not matter (they will be blended up anyway) but you are allowed repeats. So one possible shake is triple chocolate, double cherry, and mint chocolate chip.

Solution

We get six scoops, each of which could be one of ten possible flavors. Represent each scoop as a star. Think of going down the counter one flavor at a time: you see vanilla first, and skip to the next, chocolate. You say yes to chocolate three times (use three stars), then switch to the next flavor. You keep skipping until you get to cherry, which you say yes to twice. Another switch and you are at mint chocolate chip. You say yes once. Then you keep switching until you get past the last flavor, never saying yes again (since you already have said yes six times). There are ten flavors to choose from, so we must switch from considering one flavor to the next nine times. These are the nine bars.

Now that we are confident that we have the right number of stars and bars, we answer the question simply: there are 6 stars and 9 bars, so 15 symbols. We need to pick 9 of them to be bars, so the number of milkshakes possible is $\binom{15}{9}$.

Example 2.5.2

How many 7 digit phone numbers are there in which the digits are non-increasing? That is, every digit is less than or equal to the previous one.

Solution

We need to decide on 7 digits so we will use 7 stars. The bars will represent a switch from each possible single digit number down to the next smaller one. So the phone number 866-5221 is represented by the stars and bars chart

$$| * || * * | * ||| * * | * |.$$

There are 10 choices for each digit (0-9) so we must switch between choices 9 times. We have 7 stars and 9 bars, so the total number of phone numbers is $\binom{16}{9}$.

Example 2.5.3

How many integer solutions are there to the equation

$$x_1 + x_2 + x_3 + x_4 + x_5 = 13.$$

(An integer solution to an equation is a solution in which the unknown must have an integer value.)

- where $x_i \geq 0$ for each x_i?
- where $x_i > 0$ for each x_i?
- where $x_i \geq 2$ for each x_i?

Solution

This problem is just like giving 13 cookies to 5 kids. We need to say how many of the 13 units go to each of the 5 variables. In other words, we have 13 stars and 4 bars (the bars are like the "+" signs in the equation).

If x_i can be 0 or greater, we are in the standard case with no restrictions. So 13 stars and 4 bars can be arranged in $\binom{17}{4}$ ways.

Now each variable must be at least 1. So give one unit to each variable to satisfy that $\binom{12}{4}$. restriction. Now there are 8 stars left, and still 4 bars, so the number of solutions is

Now each variable must be 2 or greater. So before any counting, give each variable 2 units. We now have 3 remaining stars and 4 bars, so there are $\binom{7}{4}$ solutions.

Counting with Functions

Many of the counting problems in this section might at first appear to be examples of counting functions. After all, when we try to count the number of ways to distribute cookies to kids, we are assigning each cookie to a kid, just like you assign elements of the domain of a function to elements in the codomain. However, the number of ways to assign 7 cookies to 4 kids is $\binom{10}{7} = 120$, while the number of functions $f:\{1, 2, 3, 4, 5, 6, 7\} \to \{a, b, c, d\}$ is $4^7 = 16384$. What is going on here?

When we count functions, we consider the following two functions, for example, to be different:

$$f = \begin{pmatrix} 1 & 2 & 3 & 4 & 5 & 6 & 7 \\ a & b & c & c & c & c & c \end{pmatrix} \quad g = \begin{pmatrix} 1 & 2 & 3 & 4 & 5 & 6 & 7 \\ b & a & c & c & c & c & c \end{pmatrix}.$$

But these two functions would correspond to the *same* cookie distribution: kids a and b each get one cookie, kid c gets the rest (and none for kid d).

The point: elements of the domain are distinguished, cookies are indistinguishable. This is analogous to the distinction between permutations (like counting functions) and combinations (not).

Exercises

1. **A multiset is a collection of objects, just like a set, but can contain an object more than once (the order of the elements still doesn't matter). For example, $\{1, 1, 2, 5, 5, 7\}$ is a multiset of size 6.**
   (a) How many sets of size 5 can be made using the 10 numeric digits 0 through 9?
   (b) How many multisets of size 5 can be made using the 10 numeric digits 0 through 9?
2. **Using the digits 2 through 8, find the number of different 5-digit numbers such that:**
   (a) Digits cannot be repeated and must be written in increasing order. For example, 23678 is okay, but 32678 is not.
   (b) Digits can be repeated and must be written in *non-decreasing* order. For example, 24448 is okay, but 24484 is not.
3. **Each of the counting problems below can be solved with stars and bars. For each, say what outcome the diagram represents, if there are the correct number of stars and bars for the problem. Otherwise, say why the diagram does not represent any outcome, and what a correct diagram would look like.**

   $$* * * | * || * * |$$

   (a) How many ways are there to select a handful of 6 jellybeans from a jar that contains 5 different flavors?
   (b) How many ways can you distribute 5 identical lollipops to 6 kids?
   (c) How many 6-letter words can you make using the 5 vowels in alphabetical order?

(d) How many solutions are there to the equation $x_1 + x_2 + x_3 + x_4 = 6$.

4. **After gym class you are tasked with putting the 14 identical dodgeballs away into 5 bins.**

(a) How many ways can you do this if there are no restrictions?

(b) How many ways can you do this if each bin must contain at least one dodgeball?

5. **How many integer solutions are there to the equation $x + y + z = 8$ for which**

(a) x, y, and z are all positive?

(b) x, y, and z are all non-negative?

(c) x, y, and z are all greater than or equal to -3.

6. **When playing Yahtzee, you roll five regular 6-sided dice. How many different outcomes are possible from a single roll? The order of the dice does not matter.**

7. **Your friend tells you she has 7 coins in her hand (just pennies, nickels, dimes and quarters). If you guess how many of each kind of coin she has, she will give them to you. If you guess randomly, what is the probability that you will be correct?**

8. **How many integer solutions to $x_1 + x_2 + x_3 + x_4 = 25$ are there for which $x_1 \geq 1$, $x_2 \geq 2$, $x_3 \geq 3$ and $x_4 \geq 4$?**

9. **Solve the three counting problems below. Then say why it makes sense that they all have the same answer. That is, say how you can interpret them as each other.**

(a) How many ways are there to distribute 8 cookies to 3 kids?

(b) How many solutions in non-negative integers are there to $x + y + z = 8$?b

(c) How many different packs of 8 crayons can you make using crayons that come in red, blue and yellow?

10. **Consider functions $f: \{1, 2, 3, 4, 5\} \to \{0, 1, 2, \ldots, 9\}$.**

(a) How many of these functions are strictly increasing? Explain. (A function is strictly increasing provided if $a < b$, then $f(a) < f(b)$.)

(b) How many of the functions are non-decreasing? Explain. (A function is non-decreasing provided if $a < b$, then $f(a) \leq f(b)$.)

11. **Conic, your favorite math themed fast food drive-in offers 20 flavors which can be added to your soda. You have enough money to buy a large soda with 4 added flavors. How many different soda concoctions can you order if:**

(a) You refuse to use any of the flavors more than once?

(b) You refuse repeats but care about the order the flavors are added?

(c) You allow yourself multiple shots of the same flavor?

(d) You allow yourself multiple shots, and care about the order the flavors are added?

2.6 ADVANCED COUNTING USING PIE

Investigate!

You have 11 identical mini key-lime pies to give to 4 children. However, you don't want any kid to get more than 3 pies. How many ways can you distribute the pies?

- How many ways are there to distribute the pies without any restriction?
- Let's get rid of the ways that one or more kid gets too many pies. How many ways are there to distribute the pies if Al gets too many pies? What if Bruce gets too many? Or Cat? Or Dent?
- What if two kids get too many pies? How many ways can this happen? Does it matter which two kids you pick to overfeed?
- Is it possible that three kids get too many pies? If so, how many ways can this happen?
- How should you combine all the numbers you found above to answer the original question?

Suppose now you have 13 pies and 7 children. No child can have more than 2 pies. How many ways can you distribute the pies?

Attempt the above activity before proceeding

Stars and bars allows us to count the number of ways to distribute 10 cookies to 3 kids and natural number solutions to $x + y + z = 11$, for example. A relatively easy modification allows us to put a lower bound restriction on these problems: perhaps each kid must get at least two cookies or $x, y, z \geq 2$. This was done by first assigning each kid (or variable) 2 cookies (or units) and then distributing the rest using stars and bars.

What if we wanted an upper bound restriction? For example, we might insist that no kid gets more than 4 cookies or that $x, y, z \leq 4$. It turns out this is considerably harder, but still possible. The idea is to count all the distributions and then remove those that violate the condition. In other words, we must count the number of ways to distribute 11 cookies to 3 kids in which one or more of the kids gets more than 4 cookies. For any particular kid, this is not a problem; we do this using stars and bars. But how to combine the number of ways for kid A, or B or C? We must use the PIE.

The Principle of Inclusion/Exclusion (PIE) gives a method for finding the cardinality of the union of not necessarily disjoint sets. We saw in Section 1.1 how this works with three sets. To find how many things are in one or more of the sets A, B, and C, we should just add up the number of things in each of these sets. However, if there is any overlap among the sets, those elements are counted multiple times. So we subtract the things in each intersection of a pair of sets. But doing this removes elements which are in all three sets once too often, so we need to add it back in. In terms of cardinality of sets, we have

$$|A \cup B \cup C| = |A| + |B| + |C| - |A \cap B| - |A \cap C| - |B \cap C| + |A \cap B \cap C|.$$

Example 1.6.1

Three kids, Alberto, Bernadette, and Carlos, decide to share 11 cookies. They wonder how many ways they could split the cookies up provided that none of them receive more than 4 cookies (someone receiving no cookies is for some reason acceptable to these kids).

Solution

Without the "no more than 4" restriction, the answer would be $\binom{13}{2}$, using 11 stars and 2 bars (separating the three kids). Now count the number of ways that one or more of the kids violates the condition, i.e., gets at least 4 cookies.

Let A be the set of outcomes in which Alberto gets more than 4 cookies. Let B be the set of outcomes in which Bernadette gets more than 4 cookies. Let C be the set of outcomes in which Carlos gets more than 4 cookies. We then are looking (for the sake of subtraction) for the size of the set A ∪ B ∪ C. Using PIE, we must find the sizes of |A|, |B|, |C|, |A ∩ B| and so on. Here is what we find.

$|A| = \binom{8}{2}$. First give Alberto 5 cookies, then distribute the remaining 6 to the three kids without restrictions, using 6 stars and 2 bars.

- $|B| = \binom{8}{2}$. Just like above, only now Bernadette gets 5 cookies at the start.
- $|C| = \binom{8}{2}$. C atrlos gets 5 cookies first.
- $|A \cap B| = \binom{3}{2}$. Give Alberto and Bernadette 5 cookies each, leaving 1 (star) to distribute to the three kids (2 bars).
- $|A \cap C| = \binom{3}{2}$. Alberto and Carlos get 5 cookies first.
- $|B \cap C| = \binom{3}{2}$. Bernadette and Carlos get 5 cookies first.
- $|A \cap B \cap C| = 0$. It is not possible for all three kids to get 4 or more cookies.

Combining all of these we see

$$|A \cup B \cup C| = \binom{8}{2} + \binom{8}{2} + \binom{8}{2} - \binom{3}{2} - \binom{3}{2} - \binom{3}{2} + 0 = 75.$$

Thus the answer to the original question is $\binom{13}{2} - 75 = 78 - 75 = 3$. This makes sense now that we see it. The only way to ensure that no kid gets more than 4 cookies is to give two kids 4 cookies and one kid 3; there are three choices for which kid that should be. We could have found the answer much quicker through this observation, but the point of the example is to illustrate that PIE works!

For four or more sets, we do not write down a formula for PIE. Instead, we just think of the principle: add up all the elements in single sets, then subtract out things you counted twice (elements in the intersection of a pair of sets), then add back in elements you removed too often (elements in the intersection of groups of three sets), then take back out elements you added back in too often (elements in the intersection of groups of four sets), then add back in, take back out, add back in, etc. This would be very difficult if it wasn't for the fact that in these problems, all the cardinalities of the single sets are equal, as are all the cardinalities of the intersections of two sets, and that of three sets, and so on. Thus we can group all of these together and multiply by how many different combinations of 1, 2, 3,. . . sets there are.

Example 2.6.2

How many ways can you distribute 10 cookies to 4 kids so that no kid gets more than 2 cookies?

Solution

There are $\binom{13}{3}$ ways to distribute 10 cookies to 4 kids (using 10 stars and 3 bars). We will subtract all the outcomes in which a kid gets 3 or more cookies. How many outcomes are there like that? We can force kid A to eat 3 or more cookies by giving him 3 cookies before we start. Doing so reduces the problem to one in which we have 7 cookies to give to 4 kids without any restrictions. In that case, we have 7 stars (the 7 remaining cookies) and 3 bars (one less than the number of kids) so we can distribute the cookies in $\binom{10}{3}$ ways. Of course we could choose any one of the 4 kids to give too many cookies, so it would appear that there are $\binom{4}{1}\binom{10}{3}$ ways to distribute the cookies giving too many to one kid. But in fact, we have over counted.

We must get rid of the outcomes in which two kids have too many cookies. There are $\binom{4}{2}$ ways to select 2 kids to give extra cookies. It takes 6 cookies to do this, leaving only 4 cookies. So we have 4 stars and still 3 bars. The remaining 4 cookies can thus be distributed in $\binom{7}{3}$ ways (for each of the $\binom{4}{2}$ choices of which 2 kids to over-feed).

But now we have removed too much. We must add back in all the ways to give too many cookies to three kids. This uses 9 cookies, leaving only 1 to distribute to the 4 kids using stars and bars, which can be done in $\binom{4}{3}$ ways. We must consider this outcome for every possible choice of which three kids we over-feed, and there are $\binom{4}{3}$ ways of selecting that set of 3 kids.

Next we would subtract all the ways to give four kids too many cookies, but in this case, that number is 0.

All together we get that the number of ways to distribute 10 cookies to 4 kids without giving any kid more than 2 cookies is:

$$\binom{13}{3} - \left[\binom{4}{1}\binom{10}{3} - \binom{4}{2}\binom{7}{3} + \binom{4}{3}\binom{4}{3}\right]$$

which is

$$286 - [480 - 210 + 16] = 0.$$

This makes sense: there is NO way to distribute 10 cookies to 4 kids and make sure that nobody gets more than 2. It is slightly surprising that but since PIE works, this equality must hold.

$$\binom{13}{3} = \left[\binom{4}{1}\binom{10}{3} - \binom{4}{2}\binom{7}{3} + \binom{4}{3}\binom{4}{3}\right],$$

Just so you don't think that these problems always have easier solutions, consider the following example.

Example 2.6.3

Earlier (Example 2.5.3) we counted the number of solutions to the equation

$$x_1 + x_2 + x_3 + x_4 + x_5 = 13,$$

where $x_i \geq 0$ for each x_i.

How many of those solutions have $0 \leq x_i \leq 3$ for each x_i?

Solution

We must subtract off the number of solutions in which one or more of the variables has a value greater than 3. We will need to use PIE because counting the number of solutions for which each of the five variables separately are greater than 3 counts solutions multiple times. Here is what we get:

Total solutions: $\binom{17}{4}$.

Solutions where $x_1 > 3$: $\binom{13}{4}$. Give x1 4 units first, then distribute the remaining 9 units to the 5 variables.

Solutions where $x_1 > 3$ and $x_2 > 3$: $\binom{9}{4}$. After you give 4 units to x_1 and another 4 to x_2, you only have 5 units left to distribute.

Solutions where $x_1 > 3$, $x_2 > 3$ and $x_3 > 3$: $\binom{5}{4}$.

Solutions where $x_1 > 3$, $x_2 > 3$, $x_3 > 3$, and $x_4 > 3$: 0.

We also need to account for the fact that we could choose any of the five variables in the place of x_1 above (so there will be $\binom{5}{1}$ outcomes like this), any pair of variables in the place of x_1 and x_2 ($\binom{5}{2}$ outcomes) and so on. It is because of this that the double counting occurs, so we need to use PIE. All together we have that the number of solutions with $0 \leq x_i \leq 3$ is

$$\binom{17}{4} - \left[\binom{5}{1}\binom{13}{4} - \binom{5}{2}\binom{9}{4} + \binom{5}{3}\binom{5}{4}\right] = 15.$$

Counting Derangements

Investigate!

For your senior prank, you decide to switch the nameplates on your favorite 5 professors' doors. So that none of them feel left out, you want to make sure that all of the nameplates end up on the wrong door. How many ways can this be accomplished?

Attempt the above activity before proceeding

The advanced use of PIE has applications beyond stars and bars. A derangement of n elements $\{1, 2, 3, \ldots, n\}$ is a permutation in which no element is fixed. For example, there are 6 permutations of the three elements $\{1, 2, 3\}$:

123 132 213 231 312 321.

but most of these have one or more elements fixed: 123 has all three elements fixed since all three elements are in their original positions, 132 has the first element fixed (1 is in its original first position), and so on. In fact, the only derangements of three elements are

231 and 312.

If we go up to 4 elements, there are 24 permutations (because we have 4 choices for the first element, 3 choices for the second, 2 choices for the third leaving only 1 choice for the last). How many of these are derangements? If you list out all 24 permutations and eliminate those which are not derangements, you will be left with just 9 derangements. Let's see how we can get that number using PIE.

Example 2.6.4

How many derangements are there of 4 elements?

Solution

We count all permutations, and subtract those which are not derangements. There are 4! = 24 permutations of 4 elements. Now for a permutation to not be a derangement, at least one of the 4 elements must be fixed. There are $\binom{4}{1}$ choices for which single element we fix. Once fixed, we need to find a permutation of the other three elements. There are 3! permutations on 3 elements.

But now we have counted too many non-derangements, so we must subtract those permutations which fix two elements. There are choices $\binom{4}{2}$ for which two elements we fix, and then for each pair, 2! permutations of the remaining elements. But this subtracts too many, so add back in permutations which fix 3 elements, all $\binom{4}{3}1!$ of them. Finally subtract the $\binom{4}{4}0!$ permutations (recall 0! = 1) which fix all four elements. All together we get that the number of derangements of 4 elements is:

$$4! - \left[\binom{4}{1}3! - \binom{4}{2}2! + \binom{4}{3}1! - \binom{4}{4}0!\right] = 24 - 15 = 9.$$

Of course we can use a similar formula to count the derangements of any number of elements. However, the more elements we have, the longer the formula gets. Here is another example:

Example 2.6.5

Five gentlemen attend a party, leaving their hats at the door. At the end of the party, they hastily grab hats on their way out. How many different ways could this happen so that none of the gentlemen leave with his own hat?

Solution

We are counting derangements on 5 elements. There are 5! ways for the gentlemen to grab hats in any order—but many of these permutations will result in someone getting their own hat. So we subtract all the ways in which one or more of the men get their own hat. In other words, we subtract the non-derangements. Doing so requires PIE. Thus the answer is:

$$5! - \left[\binom{5}{1}4! - \binom{5}{2}3! + \binom{5}{3}2! - \binom{5}{4}1! + \binom{5}{5}0!\right].$$

Counting Functions

Investigate!

- **Consider all functions $f: \{1, 2, 3, 4, 5\} \to \{1, 2, 3, 4, 5\}$.** How many functions are there all together? How many of those are injective? Remember, a function is an injection if every input goes to a different output.
- **Consider all functions $f: \{1, 2, 3, 4, 5\} \to \{1, 2, 3, 4, 5\}$.** How many of the injections have the property that $f(x) \neq x$ for any $x \in \{1, 2, 3, 4, 5\}$?

Your friend claims that the answer is:

$$5! - \left[\binom{5}{1}4! - \binom{5}{2}3! + \binom{5}{3}2! - \binom{5}{4}1! + \binom{5}{5}0!\right].$$

Explain why this is correct.

- Recall that a *surjection* is a function for which every element of the codomain is in the range. How many of the functions $f: \{1, 2, 3, 4, 5\} \to \{1, 2, 3, 4, 5\}$ are surjective? Use PIE!

Attempt the above activity before proceeding

We have seen throughout this chapter that many counting questions can be rephrased as questions about counting functions with certain properties. This is reasonable since many counting questions can be thought of as counting the number of ways to assign elements from one set to elements of another.

Example 2.6.6

You decide to give away your video game collection so as to better spend your time studying advanced mathematics. How many ways can you do this, provided:

- You want to distribute your 3 different PS4 games among 5 friends, so that no friend gets more than one game?
- You want to distribute your 8 different 3DS games among 5 friends?
- You want to distribute your 8 different SNES games among 5 friends, so that each friend gets at least one game?

In each case, model the counting question as a function counting question.

Solution

We must use the three games (call them 1, 2, 3) as the domain and the 5 friends (a,b,c,d,e) as the codomain (otherwise the function would not be defined for the whole domain when a friend didn't get any game). So how many functions are there with domain $\{1, 2, 3\}$ and codomain $\{a, b, c, d, e\}$? The answer to this is $5^3 = 125$, since we can assign any of 5 elements to be the image of 1, any of 5 elements to be the image of 2 and any of 5 elements to be the image of 3.

But this is not the correct answer to our counting problem, because one of these functions is $f = \begin{pmatrix} 1 & 2 & 3 \\ a & a & a \end{pmatrix}$; one friend can get more than one game. What we really need to do is count injective functions. This gives $P(5, 3) = 60$ functions, which is the answer to our counting question.

Again, we need to use the 8 games as the domain and the 5 friends as the codomain. We are counting all functions, so the number of ways to distribute the games is 5^8.

This question is harder. Use the games as the domain and friends as the codomain (the reverse would not give a function). To ensure that every friend gets at least one game means that every element of the codomain is in the range. In other words, we are looking for surjective functions. How do you count those??

In Example 2.1.5 we saw how to count all functions (using the multiplicative principle) and in Example 1.3.4 we learned how to count injective functions (using permutations). Surjective functions are not as easily counted (unless the size of the domain is smaller than the codomain, in which case there are none).

The idea is to count the functions which are not surjective, and then subtract that from the total number of functions. This works very well when the codomain has two elements in it:

Example 2.6.7

How many functions $f: \{1, 2, 3, 4, 5\} \to \{a, b\}$ are surjective?

Solution

There are 25 functions all together, two choices for where to send each of the 5 elements of the domain. Now of these, the functions which are not surjective must exclude one or more elements of the codomain from the range. So first, consider functions for which a is not in the range. This can only happen one way: everything gets sent to b. Alternatively, we could exclude b from the range. Then everything gets sent to a, so there is only one function like this. These are the only ways in which a function could not be surjective (no function excludes both a and b from the range) so there are exactly $2^5 - 2$ surjective functions.

When there are three elements in the codomain, there are now three choices for a single element to exclude from the range. Additionally, we could pick pairs of two elements to exclude from the range, and we must make sure we don't over count these. It's PIE time!

Example 2.6.8

How many functions f: $\{1, 2, 3, 4, 5\} \to \{a, b, c\}$ are surjective?

Solution

Again start with the total number of functions: 35 (as each of the five elements of the domain can go to any of three elements of the codomain). Now we count the functions which are not surjective.

Start by excluding a from the range. Then we have two choices (b or c) for where to send each of the five elements of the domain. Thus there are 2^5 functions which exclude a from the range. Similarly, there are 2^5 functions which exclude b, and another 2^5 which exclude c. Now have we counted all functions which are not surjective? Yes, but in fact, we have counted some multiple times. For example, the function which sends everything to c was one of the 2^5 functions we counted when we excluded a from the range, and also one of the 2^5 functions we counted when we excluded b from the range. We must subtract out all the functions which specifically exclude two elements from the range. There is 1 function when we exclude a and b (everything goes to c), one function when we exclude a and c, and one function when we exclude *b* and *c*.

We are using PIE: to count the functions which are not surjective, we added up the functions which exclude *a, b,* and *c* separately, then subtracted the functions which exclude pairs of elements. We would then add back in the functions which exclude groups of three elements, except that there are no such functions. We find that the number of functions which are not surjective is

$$2^5 + 2^5 + 2^5 - 1 - 1 - 1 + 0.$$

Perhaps a more descriptive way to write this is since each of the 2^5's was the result of choosing 1 of the 3 elements of the codomain to exclude from the range, each of the three 1^5's was the result of choosing 2 of the 3 elements of the codomain to exclude. Writing 1^5 instead of 1 makes sense too: we have 1 choice of where to send each of the 5 elements of the domain.

$$\binom{3}{1}2^5 - \binom{3}{2}1^5 + \binom{3}{3}0^5.$$

Now we can finally count the number of surjective functions:

$$3^5 - \left[\binom{3}{1}2^5 - \binom{3}{2}1^5\right] = 150.$$

You might worry that to count surjective functions when the codomain is larger than 3 elements would be too tedious. We need to use PIE but with more than 3 sets the formula for PIE is very long. However, we have lucked out. As we saw in the example above, the number of functions which exclude a single element from the range is the same no matter which single element is excluded. Similarly, the number of functions which exclude a pair of elements will be the same for every pair. With larger codomains, we will see the same behavior with groups of 3, 4, and more elements excluded. So instead of adding/subtracting each of these, we can simply add or subtract all of them at once, if you know how many there are. This works just like it did in for the other types of counting questions in this section, only now the size of the various combinations of sets is a number raised to a power, as opposed to a binomial coefficient or factorial. Here's what happens with 4 and 5 elements in the codomain.

Example 2.6.9

- **How many functions f:** $\{1, 2, 3, 4, 5\} \to \{a, b, c, d\}$ are surjective?
- **How many functions f:** $\{1, 2, 3, 4, 5\} \to \{a, b, c, d, e\}$ are surjective?

Solution

There are 4^5 functions all together; we will subtract the functions which are not surjective. We could exclude any one of the four elements of the codomain, and doing so will leave us with 3^5 functions for each excluded element. This counts too many so we subtract the functions which exclude two of the four elements of the codomain, each pair giving 2^5 functions. But this excludes too many, so we add back in the functions which exclude three of the four elements of the codomain, each triple giving 1^5 function. There are $\binom{4}{1}$ groups of functions excluding a single element, $\binom{4}{2}$ groups of functions excluding a pair of elements, and $\binom{4}{3}$ groups of functions excluding a triple of elements. This means that the number of functions which are not surjective is:

$$\binom{4}{1}3^5 - \binom{4}{2}2^5 + \binom{4}{3}1^5.$$

We can now say that the number of functions which are surjective is:

$$4^5 - \left[\binom{4}{1}3^5 - \binom{4}{2}2^5 + \binom{4}{3}1^5\right].$$

The number of surjective functions is:

$$5^5 - \left[\binom{5}{1}4^5 - \binom{5}{2}3^5 + \binom{5}{3}2^5 - \binom{5}{4}1^5\right].$$

We took the total number of functions 5^5 and subtracted all that were not surjective. There were $\binom{5}{1}$ ways to select a single element from the codomain to exclude from the range, and for each there were 4^5 functions. But this double counts, so we use PIE and subtract functions excluding two elements from the range: there are $\binom{5}{2}$ choices for the two elements to exclude, and for each pair, 3^5 functions. This takes out too many functions, so we add back in functions which exclude 3 elements from the range: $\binom{5}{3}$ choices for which three to exclude, and then 2^5 functions for each choice of elements. Finally we take back out the 1 function which excludes 4 elements for each of the $\binom{5}{4}$ choices of 4 elements.

If you happen to calculate this number precisely, you will get 120 surjections. That happens to also be the value of 5!. This might seem like an amazing coincidence until you realize that every surjective function $f: X \to Y$ with $|X| = |Y|$ finite must necessarily be a bijection. The number of bijections is always $|X|!$ in this case. What we have here is a combinatorial proof of the following identity:

$$n^n - \left[\binom{n}{1}(n-1)^n - \binom{n}{2}(n-2)^n + \cdots + \binom{n}{n-1}1^n\right] = n!.$$

We have seen that counting surjective functions is another nice example of the advanced use of the Principle of Inclusion/Exclusion. Also, counting injective functions turns out to be equivalent to permutations, and counting all functions has a solution akin to those counting problems where order matters but repeats are allowed (like counting the number of words you can make from a given set of letters).

These are not just a few more examples of the techniques we have developed in this chapter. Quite the opposite: everything we have learned in this chapter are examples of *counting functions!*

Example 2.6.10

How many 5-letter words can you make using the eight letters a through h? How many contain no repeated letters?

Solution

By now it should be no surprise that there are 85 words, and P 8, 5 words without repeated letters. The new piece here is that we are actually counting functions. For the first problem, we are counting all functions from $\{1, 2, \ldots, 5\}$ to $\{a, b, \ldots, h\}$. The numbers in the domain represent the *position* of the letter in the word, the codomain represents the letter that could be assigned to that position. If we ask for no repeated letters, we are asking for injective functions.

If A and B are *any* sets with $|A| = 5$ and $|B| = 8$, then the number of functions $f : A \to B$ is 8^5 and the number of injections is $P(8, 5)$. So if you can represent your counting problem as a function counting problem, most of the work is done.

Example 2.6.11

How many subsets are there of $\{1, 2, \ldots, 9\}$? How many 9-bit strings are there (of any weight)?

Solution

We saw in Section 1.2 that the answer to both these questions is 2^9, as we can say yes or no (or 0 or 1) to each of the 9 elements in the set (positions in the bit-string). But 29 also looks like the answer you get from counting functions. In fact, if you count all functions $f : A \to B$ with $|A| = 9$ and $|B| = 2$, this is exactly what you get.

This makes sense! Let $A = \{1, 2, \ldots, 9\}$ and $B = \{y, n\}$. We are assigning each element of the set either a yes or a no. Or in the language of bit-strings, we would take the 9 positions in the bit string as our domain and the set $\{0, 1\}$ as the codomain.

So far we have not used a function as a model for binomial coefficients (combinations). Think for a moment about the relationship between combinations and permutations, say specifically $\binom{9}{3}$ and $P(9, 3)$. We do have a function model for $P(9, 3)$. This is the number of injective functions from a set of size 3 (say $\{1, 2, 3\}$ to a set of size 9 (say $\{1, 2, \ldots, 9\}$) since there are 9 choices for where to send the first element of the domain, then only 8 choices for the second, and 7 choices for the third. For example, the function might look like this:

$$f(1) = 5 \qquad f(2) = 8 \qquad f(3) = 4.$$

This is a different function from:

$$f(1) = 4 \qquad f(2) = 5 \qquad f(3) = 8.$$

Now $P(9, 3)$ counts these as different outcomes correctly, but $\binom{9}{3}$ will count these (among others) as just one outcome. In fact, in terms of functions $\binom{9}{3}$ just counts the number of different ranges possible of injective functions. This should not be a surprise since binomial coefficients counts subsets, and the range is a possible subset of the codomain.

While it is possible to interpret combinations as functions, perhaps the better advice is to instead use combinations (or stars and bars) when functions are not quite the right way to interpret the counting question.

Exercises

1. **The dollar menu at your favorite tax-free fast food restaurant has 7 items. You have $10 to spend. How many different meals can you buy if you spend all your money and:**
   (a) Purchase at least one of each item.
   (b) Possibly skip some items.
   (c) Don't get more than 2 of any particular item.
2. **After a late night of math studying, you and your friends decide to go to your favorite tax-free fast food Mexican restaurant, Burrito Chime. You decide to order off of the dollar menu, which has 7 items. Your group has $16 to spend (and will spend all of it).**
   (a) How many different orders are possible? Explain. (The order in which the order is placed does not matter - just which and how many of each item that is ordered.)
   (b) How many different orders are possible if you want to get at least one of each item? Explain.
   (c) How many different orders are possible if you don't get more than 4 of any one item? Explain.

3. **After another gym class you are tasked with putting the 14 identical dodgeballs away into 5 bins. This time, no bin can hold more than 6 balls. How many ways can you clean up?**

4. **Consider the equation $x_1 + x_2 + x_3 + x_4 = 15$. How many solutions are there with $2 \le x_i \le 5$ for all $i \in \{1, 2, 3, 4\}$?**

5. **Suppose you planned on giving 7 gold stars to some of the 13 star students in your class. Each student can receive at most one star. How many ways can you do this?**

Use PIE. Then, find the numeric answer in Pascal's triangle and explain why that makes sense.

6. **Based on the previous question, give a combinatorial proof for the identity:**

$$\binom{n}{k} = \binom{n+k-1}{k} - \sum_{j=1}^{n}(-1)^{j+1}\binom{n}{j}\binom{n+k-(2j+1)}{k-2j}.$$

7. **Illustrate how the counting of derangements works by writing all permutations of {1, 2, 3, 4} and the crossing out those which are not derangements. Keep track of the permutations you cross out more than once, using PIE.**

8. **How many permutations of {1, 2, 3, 4, 5} leave exactly 1 element fixed?**

9. **Ten ladies of a certain age drop off their red hats at the hat check of a museum. As they are leaving, the hat check attendant gives the hats back randomly. In how many ways can exactly six of the ladies receive their own hat (and the other four not)? Explain.**

10. **The Grinch sneaks into a room with 6 Christmas presents to 6 different people. He proceeds to switch the name-labels on the presents. How many ways could he do this if:**

    (a) No present is allowed to end up with its original label? Explain what each term in your answer represents.

    (b) Exactly 2 presents keep their original labels? Explain.

    (c) Exactly 5 presents keep their original labels? Explain.

11. **Consider functions $f: \{1, 2, 3, 4\} \to \{a, b, c, d, e, f\}$. How many functions have the property that $f(1) \neq a$ or $f(2) \neq b$, or both?**

12. **Consider sets A and B with $|A| = 10$ and $|B| = 5$. How many functions $f: A \to B$ are surjective?**

13. **Let $A = \{1, 2, 3, 4, 5\}$. How many injective functions $f: A \to A$ have the property that for each $x \in A$, $f(x) \neq x$?**

14. Let d_n be the number of derangements of n objects. For example, using the techniques of this section, we find

$$d_3 = 3! - \left(\binom{3}{1}2! - \binom{3}{2}1! + \binom{3}{3}0!\right).$$

We can use the formula for $\binom{n}{k}$ to write this all in terms of factorials. After simplifying, for d3 we would get

$$d_3 = 3!\left(1 - \frac{1}{1} + \frac{1}{2} - \frac{1}{6}\right).$$

Generalize this to find a nicer formula for d_n. Bonus: For large n, approximately what fraction of all permutations are derangements? Use your knowledge of Taylor series from calculus.

2.7 CONCISE SUMMARY

Investigate!

Suppose you have a huge box of animal crackers containing plenty of each of 10 different animals. For the counting questions below, carefully examine their similarities and differences, and then give an answer. The answers are all one of the following:

$$P(10,6) \qquad \binom{10}{6} \qquad 10^6 \qquad \binom{15}{9}.$$

- How many animal parades containing 6 crackers can you line up?
- How many animal parades of 6 crackers can you line up so that the animals appear in alphabetical order?
- How many ways could you line up 6 different animals in alphabetical order?
- How many ways could you line up 6 different animals if they can come in any order?
- How many ways could you give 6 children one animal cracker each?
- How many ways could you give 6 children one animal cracker each so that no two kids get the same animal?
- How many ways could you give out 6 giraffes to 10 kids?
- Write a question about giving animal crackers to kids that has the answer $\binom{10}{6}$.

Attempt the above activity before proceeding

With all the different counting techniques we have mastered in this last chapter, it might be difficult to know when to apply which technique. Indeed, it is very easy to get mixed up and use the wrong counting method for a given problem. You get better with practice. As you practice you start to notice some trends that can help you distinguish between types of counting problems. Here are some suggestions that you might find helpful when deciding how to tackle a counting problem and checking whether your solution is correct.

Remember that you are counting the number of items in some *list of outcomes.* Write down part of this list. Write down an element in the middle of the list – how are you deciding whether your element really is in the list. Could you get this element more than once using your proposed answer?

If generating an element on the list involves selecting something (for example, picking a letter or picking a position to put a letter, etc), can the things you select be repeated? Remember, permutations and combinations select objects from a set *without* repeats.

Does order matter? Be careful here and be sure you know what your answer really means. We usually say that order matters when you get different outcomes when the same objects are selected in different orders. Combinations and "Stars & Bars" are used when order *does not* matter.

There are four possibilities when it comes to order and repeats. If order matters and repeats are allowed, the answer will look like n^k. If order matters and repeats are not allowed, we have $P(n, k)$. If order doesn't matter and repeats are allowed, use stars and bars. If order doesn't matter and repeats are not allowed, use $\binom{n}{k}$. But be careful: this only applies when you are selecting things, and you should make sure you know exactly what you are selecting before determining which case you are in.

Think about how you would represent your counting problem in terms of sets or functions. We know how to count different sorts of sets and different types of functions.

As we saw with combinatorial proofs, you can often solve a counting problem in more than one way. Do that, and compare your numerical answers. If they don't match, something is amiss.

While we have covered many counting techniques, we have really only scratched the surface of the large subject of *enumerative combinatorics.* There are mathematicians doing original research in this area even as you read this. Counting can be really hard.

In the next chapter, we will approach counting questions from a very different direction, and in doing so, answer infinitely many counting questions at the same time. We will create *sequences* of answers to related questions.

Chapter Review

1. **You have 9 presents to give to your 4 kids. How many ways can this be done if:**
   (a) The presents are identical, and each kid gets at least one present?
   (b) The presents are identical, and some kids might get no presents?
   (c) The presents are unique, and some kids might get no presents?
   (d) The presents are unique and each kid gets at least one present?

2. **For each of the following counting problems, say whether the answer is $\binom{10}{4}$, $P(10,4)$, or neither. If you answer is "neither," say what the answer should be instead.**
   (a) How many shortest lattice paths are there from (0, 0) to (10, 4)?
   (b) If you have 10 bow ties, and you want to select 4 of them for next week, how many choices do you have?
   (c) Suppose you have 10 bow ties and you will wear a different one on each of the next 4 days. How many choices do you have?
   (d) If you want to wear 4 of your 10 bow ties next week (Monday through Sunday), how many ways can this be accomplished?
   (e) Out of a group of 10 classmates, how many ways can you rank your top 4 friends?
   (f) If 10 students come to their professor's office but only 4 can fit at a time, how different combinations of 4 students can see the prof first?
   (g) How many 4 letter words can be made from the first 10 letters of the alphabet?
   (h) How many ways can you make the word "cake" from the first 10 letters of the alphabet?
   (i) How many ways are there to distribute 10 identical apples among 4 children?
   (j) If you have 10 kids (and live in a shoe) and 4 types of cereal, how many ways can your kids eat breakfast?
   (k) How many ways can you arrange exactly 4 ones in a string of 10 binary digits?
   (l) You want to select 4 distinct, single-digit numbers as your lotto picks. How many choices do you have?
   (m) 10 kids want ice-cream. You have 4 varieties. How many ways are there to give the kids as much ice-cream as they want?
   (n) How many 1-1 functions are there from $\{1, 2, \ldots, 10\}$ to $\{a, b, c, d\}$?

(o) How many surjective functions are there from {1, 2, . . . , 10} to {a, b, c, d}?

(p) Each of your 10 bow ties match 4 pairs of suspenders. How many outfits can you make?

(q) After the party, the 10 kids each choose one of 4 party-favors. How many outcomes?

(r) How many 6-elements subsets are there of the set {1, 2, . . . , 10}

(s) How many ways can you split up 11 kids into 5 named teams?

(t) How many solutions are there to $x_1 + x_2 + \ldots\ldots + x_5 = 6$ where each xi is a non-negative integer?

(u) Your band goes on tour. There are 10 cities within driving distance, but only enough time to play 4 of them. How many choices do you have for the cities on your tour?

(v) In how many different ways can you play the 4 cities you choose?

(w) Out of the 10 breakfast cereals available, you want to have 4 bowls. How many ways can you do this?

(x) There are 10 types of cookies available. You want to make a 4 cookie stack. How many different stacks can you make?

(y) From your home at (0,0) you want to go to either the donut shop at (5,4) or the one at (3,6). How many paths could you take?

(z) How many 10-digit numbers do not contain a sub-string of 4 repeated digits?

3. **Bow tiesRecall, you own 3 regular ties and 5 bow ties. You realize that it would be okay to wear more than two ties to your clown college interview.**

(a) You must select some of your ties to wear. Everything is okay, from no ties up to all ties. How many choices do you have?

(b) If you want to wear at least one regular tie and one bow tie, but are willing to wear up to all your ties, how many choices do you have for which ties to wear?

(c) How many choices of which ties to wear do you have if you wear exactly 2 of the 3 regular ties and 3 of the 5 bow ties?

(d) Once you have selected 2 regular and 3 bow ties, in how many orders could you put the ties on, assuming you must have one of the three bow ties on top?

4. **Give a counting question where the answer is 8. 3. 3. 5. Give another question where the answer is 8 + 3 + 3 + 5.**

5. **Consider five digit numbers $\alpha = a_1a_2a_3a_4a_5$, with each digit from the set {1, 2, 3, 4}.**

(a) How many such numbers are there?

(b) How many such numbers are there for which the sum of the digits is even?

(c) How many such numbers contain more even digits than odd digits?

6. **In a recent small survey of airline passengers, 25 said they had flown American in the last year, 30 had flown Jet Blue, and 20 had flown Continental. Of those, 10 reported they had flown on American and Jet Blue, 12 had flown on Jet Blue and Continental, and 7 had flown on American and Continental. 5 passengers had flown on all three airlines.**

How many passengers were surveyed? (Assume the results above make up the entire survey.)

7. **Recall, by 8-bit strings, we mean strings of binary digits, of length 8.**

(a) How many 8-bit strings are there total?

(b) How many 8-bit strings have weight 5?

(c) How many subsets of the set $\{a, b, c, d, e, f, g, h\}$ contain exactly 5 elements?

(d) Explain why your answers to parts (*b*) and (*c*) are the same. Why are these questions equivalent?

8. **What is the coefficient of x10 in the expansion of $(x + 1)^{13} + x^2(x + 1)^{17}$?**

9. **How many 8-letter words contain exactly 5 vowels? (One such word is "aaioobtt"; don't consider "y" a vowel for this exercise.) What if repeated letters were not allowed?**

10. **For each of the following, find the number of shortest lattice paths from (0, 0) to (8, 8) which:**

(a) pass through the point (2, 3).

(b) avoid (do not pass through) the point (7, 5).

(c) either pass through (2, 3) or (5, 7) (or both).

11. **You live in Grid-Town on the corner of 2nd and 3rd, and work in a building on the corner of 10th and 13th. How many routes are there which take you from home to work and then back home, but by a different route?**

12. **How many 10-bit strings start with 111 or end with 101 or both?**

13. **How many 10-bit strings of weight 6 start with 111 or end with 101 or both?**

14. **How many 6 letter words made from the letters a, b, c, d, e, f without repeats do not contain the sub-word "bad" in consecutive letters?**

15. **How many don't to contain the subword "bad" in not-necessarily consecutive letters (but in order)?**

Explain using lattice paths why $\sum_{k=0}^{n} \binom{n}{k} = 2^n$.

16. **Suppose you have 20 one-dollar bills to give out as prizes to your top 5 discrete math students. How many ways can you do this if:**

(a) Each of the 5 students gets at least 1 dollar?

(b) Some students might get nothing?

(c) Each student gets at least 1 dollar but no more than 7 dollars?

17. How many functions $f : \{1, 2, 3, 4, 5\} \rightarrow \{a, b, c, d, e\}$ are there satisfying:

(a) $f(1) = a$ or $f(2) = b$ (or both)?

(b) $f(1) \neq a$ or $f(2) * b$ (or both)?

(c) $f(1) \neq a$ and $f(2) * b$, and f is injective?

(d) f is surjective, but $f(1) \neq a$, $f(2) \neq b$, $f(3) \neq c$, $f(4) \neq d$ and $f(5) \neq e$?

18. How many functions map {1, 2, 3, 4, 5, 6} *onto* {*a*, *b*, *c*, *d*} (i.e., how many *surjections* are there)?

19. To thank your math professor for doing such an amazing job all semester, you decide to bake Oscar cookies. You know how to make 10 different types of cookies.

(a) If you want to give your professor 4 different types of cookies, how many different combinations of cookie type can you select? Explain your answer.

(b) To keep things interesting, you decide to make a different number of each type of cookie. If again you want to select 4 cookie types, how many ways can you select the cookie types and decide for which there will be the most, second most, etc. Explain your answer.

(c) You change your mind again. This time you decide you will make a total of 12 cookies. Each cookie could be any one of the 10 types of cookies you know how to bake (and it's okay if you leave some types out). How manychoices do you have? Explain.

(d) You realize that the previous plan did not account for presentation. This time, you once again want to make 12 cookies, each one could be any one of the 10 types of cookies. However, now you plan to shape the cookies into the numerals 1, 2, ..., 12 (and probably arrange them to make a giant clock, but you haven't decided on that yet). How many choices do you have for which types of cookies to bake into which numerals? Explain.

(e) The only flaw with the last plan is that your professor might not get to sample all 10 different varieties of cookies. How many choices do you have for which types of cookies to make into which numerals, given that each type of cookie should be present at least once? Explain.

For which of the parts of the previous problem (Exercise 1.7.19) does it make sense to interpret the counting question as counting some number of functions? Say what the domain and codomain should be, and whether you are counting all functions, injections, surjections, or something else.

FOOTNOTES

1. To solve this problem, you could break it into two cases. First, count how many ways there are to select the two cards when the first card is a red non-face card. Second, count how many ways when the first card is a red face card. Doing so makes the events in each separate case independent, so the multiplicative principle can be applied.
2. Most every binomial identity can be proved using mathematical induction, using the recursive definition for $\binom{n}{k}$. We will discuss induction in Section 2.5.
3. A more mathematically sophisticated interpretation of combinations is that we are defining two injective functions to be *equivalent* if they have the same range, and then counting the number of equivalence classes under this notion of equivalence.

CHAPTER-3

SEQUENCES

INTRODUCTION

In mathematics, sequences represent ordered lists of numbers, each following a specific pattern or rule. From the simple arithmetic progression to the intricacies of Fibonacci or geometric sequences, these mathematical structures provide a framework for understanding and predicting numerical patterns. Sequences are foundational in numerous branches of mathematics and beyond, influencing fields such as computer science, economics, and physics. They offer insights into growth patterns, recurrence relations, and even the behavior of complex systems. This exploration delves into the diversity and significance of sequences, examining their historical development, theoretical underpinnings, practical applications, and their role in shaping our comprehension of mathematical principles and natural phenomena.

Investigate!

There is a monastery in Hanoi, as the legend goes, with a great hall containing three tall pillars. Resting on the first pillar are 64 giant disks (or washers), all different sizes, stacked from largest to smallest. The monks are charged with the following task: they must move the entire stack of disks to the third pillar. However, due to the size of the disks, the monks cannot move more than one at a time. Each disk must be placed on one of the pillars before the next disk is moved. And because the disks are so heavy and fragile, the monks may never place a larger disk on top of a smaller disk. When the monks finally complete their task, the world shall come to an end. Your task: figure out how long before we need to start worrying about the end of the world.

- First, let's find the minimum number of moves required for a smaller number of disks. Collect some data. Make a table.
- Conjecture a formula for the minimum number of moves required to move n disks. Test your conjecture. How do you know your formula is correct?

- If the monks were able to move one disk every second without ever stopping, how long before the world ends?

Attempt the above activity before proceeding

This puzzle is called the *Tower of Hanoi.* You are tasked with finding the minimum number of moves to complete the puzzle. This certainly sounds like a counting problem. Perhaps you have an answer? If not, what else could we try?

The answer depends on the number of disks you need to move. In fact, we could answer the puzzle first for 1 disk, then 2, then 3 and so on. If we list out all of the answers for each number of disks, we will get a **sequence** of numbers. The nth term in the sequence is the answer to the question, "what is the smallest number of moves required to complete the Tower of Hanoi puzzle with n disks?" You might wonder why we would create such a sequence instead of just answering the question. By looking at how the sequence of numbers grows, we gain insight into the problem. It is easy to count the number of moves required for a small number of disks. We can then look for a pattern among the first few terms of the sequence. Hopefully this will suggest a method for finding the nth term of the sequence, which is the answer to our question. Of course we will also need to verify that our suspected pattern is correct, and that this correct pattern really does give us the nth term we think it does, but it is impossible to prove that your formula is correct without having a formula to start with.

Sequences are also interesting mathematical objects to study in their own right. Let's see why.

3.1 DESCRIBING SEQUENCES

Investigate!

You have a large collection of 1 × 1 squares and 1 × 2 dominoes. You want to arrange these to make a 1 × 15 strip. How many ways can you do this?

- Start by collecting data. How many length 1 × 1 strips can you make? How many 1 × 2 strips? How many 1 × 3 strips? And so on.
- How are the 1 × 3 and 1 × 4 strips related to the 1 × 5 strips?
- How many 1 × 15 strips can you make?
- What if I asked you to find the number of 1 × 1000 strips? Would the method you used to calculate the number fo 1 × 15 strips be helpful?

Attempt the above activity before proceeding

A sequence is simply an ordered list of numbers. For example, here is a sequence: 0, 1, 2, 3, 4, 5,...... This is different from the set ℕ because, while the sequence is a complete list of every element in the set of natural numbers, in the sequence we very much care what order the numbers come in. For this reason, when we use variables to represent terms in a sequence they will look like this:

$$a_0, a_1, a_2, a_3, \ldots\ldots\ldots$$

To refer to the entire sequence at once, we will write $(an)_{n\in\mathbb{N}}$ or $(a_n)_{n\geq 0}$, or sometimes if we are being sloppy, just (a_n) (in which case we assume we start the sequence with a_0).

We might replace the *a* with another letter, and sometimes we omit a_0, starting with a_1, in which case we would use an n≥1 to refer to the sequence as a whole. The numbers in the subscripts are called **indices** (the plural of **index**).

While we often just think of a sequence as an ordered list of numbers, it is really a type of function. Specifically, the sequence $(a_n)_{n\geq 0}$ is a function with domain $\mathbb{N}$ where a_n is the image of the natural number n. Later we will manipulate sequences in much the same way you have manipulated functions in algebra or calculus. We can shift a sequence up or down, add two sequences, or ask for the rate of change of a sequence. These are done exactly as you would for functions.

That said, while keeping the rigorous mathematical definition in mind is helpful, we often describe sequences by writing out the first few terms.

Example 3.1.1

Can you find the next term in the following sequences?

1. 7, 7, 7, 7, 7, . . .
2. 3, −3, 3, −3, 3, . . .
3. 1, 5, 2, 10, 3, 15, . . .
4. 1, 2, 4, 8, 16, 32, . . .
5. 1, 4, 9, 16, 25, 36, . . .
6. 1, 2, 3, 5, 8, 13, 21, . . .
7. 1, 3, 6, 10, 15, 21, . . .
8. 2, 3, 5, 7, 11, 13, . . .
9. 3, 2, 1, 0, −1, . . .
10. 1, 1, 2, 6, . . .

Solution. No you cannot. You might guess that the next terms are:

1. 7
2. −3
3. 4

4. 64
5. 49
6. 34
7. 28
8. 17
9. -2
10. 24

In fact, those are the next terms of the sequences I had in mind when I made up the example, but there is no way to be sure they are correct.

Still, we will often do this. Given the first few terms of a sequence, we can ask what the pattern in the sequence suggests the next terms are.

Given that no number of initial terms in a sequence is enough to say for certain which sequence we are dealing with, we need to find another way to specify a sequence. We consider two ways to do this:

Closed formula

A **closed formula** for a sequence $(a_n)_{n\in\mathbb{N}}$ is a formula for an using a fixed finite number of operations on *n*. This is what you normally think of as a formula in *n*, just as if you were defining a function in terms of n (because that is exactly what you are doing).

Recursive definition

A **recursive definition** (sometimes called an **inductive definition**) for a sequence $(a_n)_{n\in\mathbb{N}}$ consists of a **recurrence relation** : an equation relating a term of the sequence to previous terms (terms with smaller index) and an **initial condition:** a list of a few terms of the sequence (one less than the number of terms in the recurrence relation).

It is easier to understand what is going on here with an example:

Example 3.1.2

Here are a few closed formulas for sequences: $a_n = n^2$.

$$a_n = \frac{n(n+1)}{2}.$$

$$a_n = \frac{\left(\frac{1+\sqrt{5}}{2}\right)^n - \left(\frac{1-\sqrt{5}}{2}\right)^{-n}}{\sqrt{5}}.$$

Note in each formula, if you are given *n*, you can calculate a_n directly: just plug in *n*. For example, to find a_3 in the second sequence, just compute $a_3 = \frac{3(3+1)}{2} = 6$.

Here are a few recursive definitions for sequences:

- $a_n = 2a_{n-1}$ with $a_0 = 1$.
- $a_n = 2a_{n-1}$ with $a_0 = 27$.
- $a_n = a_{n-1}$ + an-2 with $a_0 = 0$ and $a_1 = 1$.

In these formulas, if you are given n, you cannot calculate a_n directly, you first need to find a_{n-1} (or a_{n-1} and a_{n-2}). In the second sequence, to find a_3 you would take $2a_2$, but to find $a_2 = 2a_1$ we would need to know $a_1 = 2a_0$. We do know this, so we could trace back through these equations to find $a_1 = 54$, $a_2 = 108$ and finally $a_3 = 216$.

You might wonder why we would bother with recursive definitions for sequences. After all, it is harder to find a_n with a recursive definition than with a closed formula. This is true, but it is also harder to find a closed formula for a sequence than it is to find a recursive definition. So to find a useful closed formula, we might first find the recursive definition, then use that to find the closed formula.

This is not to say that recursive definitions aren't useful in finding an. You can always calculate a_n given a recursive definition, it might just take a while.

Example 3.1.3

Find a_6 in the sequence defined by $a_n = 2a_n-1\ a_{n-2}$ with $a_0 = 3$ and $a_1 = 4$.

Solution. We know that $a_6 = 2a_5\ a_4$. So to find a_6 we need to find a_5 and a_4. Well

$$a_5 = 2a_4 - a_3 \qquad \text{and} \qquad a_4 = 2a_3 - a_2,$$

so if we can only find a_3 and a_2 we would be set. Of course

$$a_3 = 2a_2 - a_1 \qquad \text{and} \qquad a_2 = 2a_1 - a_0,$$

so we only need to find a_1 and a_0. But we are given these. Thus

$$\begin{aligned}
a_0 &= 3\\
a_1 &= 4\\
a_2 &= 2 \cdot 4 - 3 = 5\\
a_3 &= 2 \cdot 5 - 4 = 6\\
a_4 &= 2 \cdot 6 - 5 = 7\\
a_5 &= 2 \cdot 7 - 6 = 8\\
a_6 &= 2 \cdot 8 - 7 = 9.
\end{aligned}$$

Note that now we can guess a closed formula for the nth term of the sequence: $a_n = n + 3$. To be sure this will always work, we could plug in this formula into the recurrence relation:

$$\begin{aligned} 2a_{n-1} - a_{n-2} &= 2((n-1)+3) - ((n-2)+3) \\ &= 2n + 4 - n - 1 \\ &= n + 3 = a_n. \end{aligned}$$

That is not quite enough though, since there can be multiple closed formulas that satisfy the same recurrence relation; we must also check that our closed formula agrees on the initial terms of the sequence. Since $a_0 = 0 + 3 = 3$ and $a_1 = 1 + 3 = 4$ are the correct initial conditions, we can now conclude we have the correct closed formula.

Finding closed formulas, or even recursive definitions, for sequences is not trivial. There is no one method for doing this. Just as in evaluating integrals or solving differential equations, it is useful to have a bag of tricks you can apply, but sometimes there is no easy answer.

Common Sequences.

$$1, 4, 9, 16, 25, \ldots$$

The **square numbers.** The sequence $(s_n)_{n\geq 1}$ has closed formula $s_n = n^2$

$$1, 3, 6, 10, 15, 21, \ldots$$

The **triangular numbers.** The sequence $(T_n)_{n\geq 1}$ has closed formula $T_n = \frac{n(n+1)}{2}$.

$$1, 2, 4, 8, 16, 32, \ldots$$

The powers of 2. The sequence $(a_n)_{n\geq 0}$ with closed formula $a_n=2^n$.

$$1, 1, 2, 3, 5, 8, 13, \ldots.$$

The Fibonacci numbers (or Fibonacci sequence), defined recursively by $F_n = F_{n-1} + F_{n-2}$ with $F_1 = F_2 = 1$.

Example 3.1.4

Use the formulas $T_n = \frac{n(n+1)}{2}$ and $a_n = 2^n$ to find closed formulas that agree with the following sequences. Assume each first term corresponds to $n = 0$.

1. (b_n): 1, 2, 4, 7, 11, 16, 22, ...
2. (c_n): 3, 5, 9, 17, 33, ...
3. (d_n): 0, 2, 6, 12, 20, 30, 42, ...
4. (e_n): 3, 6, 10, 15, 21, 28, ...
5. (f_n): 0, 1, 3, 7, 15, 31, ...
6. (g_n) 3, 6, 12, 24, 48, ...
7. (h_n): 6, 10, 18, 34, 66, ...
8. (j_n): 15, 33, 57, 87, 123, ...

Solution

We wish to compare these sequences to the triangular numbers (0, 1, 3, 6, 10, 15, 21,), when we start with n = 0, and the powers of 2: (1, 2, 4, 8, 16,). (1, 2, 4, 7, 11, 16, 22, ...) Note that if subtract 1 from each term, we get the sequence T_n . So we have $b_n = T_n + 1$. Therefore a closed formula is $b_n = \frac{n(n+1)}{2} + 1$. A quick check of the first few n confirms we have it right.

(3, 5, 9, 17, 33, ...) Each term in this sequence is one more than a power of 2, so we might guess the closed formula is $c_n = a_n + 1 = 2^n + 1$. If we try this though, we get $c_0 = 2^0 + 1 = 2$ and $c^1 = 2^1 + 1 = 3$. We are off because the indices are shifted. What we really want is cn = $a_{n+1} + 1$ giving $c_n = 2^{n+1} + 1$.

(0, 2, 6, 12, 20, 30, 42, ...) Notice that all these terms are even. What happens if we factor out a 2? We get (T_n)! More precisely, we find that $d_n/2 = T_n$, so this sequence has closed formula $d_n = n(n + 1)$.

(3, 6, 10, 15, 21, 28, ...). These are all triangular numbers. However, we are starting with 3 as our initial term instead of as our third term. So if we could plug in 2 instead of 0 into the formula for T_n, we would be set. Therefore the closed formula is $e_n = \frac{(n+2)(n+3)}{2}$ (where n + 3 came from n + 2 + 1). Thinking about sequences as functions, we are doing a horizontal shift by 2: $e_n = T_{n+2}$ which would cause the graph to shift 2 units to the left.

(0, 1, 3, 7, 15, 31, ...). Try adding 1 to each term and we get powers of 2. You might guess this because each term is a little more than twice the previous term (the powers of 2 are exactly twice the previous term). Closed formula: $f_n = 2^n - 1$.

(3, 6, 12, 24, 48, ...). These numbers are also doubling each time, but are also all multiples of 3. Dividing each by 3 gives 1, 2, 4, 8,......... Aha. We get the closed formula $g_n = 3 \cdot 2^n$.

(6, 10, 18, 34, 66, . . .). To get from one term to the next, we almost double each term. So maybe we can relate this back to 2^n . Yes, each term is 2 more than a power of 2. So we get $h_n = 2^{n+2} + 2$ (the n + 2 is because the first term is 2 more than 2^2, not 2^0). Alternatively, we could have related this sequence to the second sequence in this example: starting with 3, 5, 9, 17, . . . we see that this sequence is twice the terms from that sequence. That sequence had closed formula $c_n = 2^{n+1} + 1$.

Our sequence here would be twice this, so $h_n = 2(2^n + 1)$, which is the same as we got before. (15, 33, 57, 87, 123, . . .). Try dividing each term by 3. That gives the sequence 5, 11, 19, 29, 41, Now add 1 to each term: 6, 12, 20, 30, 42, . . ., which is (d_n) in this example, except starting with 6 instead of 0. So let's start with the formula $d_n = n(n + 1)$. To start with the 6, we shift: $(n + 2)(n + 3)$. But this is one too many, so subtract 1: $(n + 2)(n + 3) - 1$. That gives us our sequence, but divided by 3. So we want $j_n = 3((\underline{n} + 2)(n + 3) - 1)$.

Partial sums

Some sequences naturally arise as the sum of terms of another sequence.

Example 3.1.5

Sam keeps track of how many push-ups she does each day of her "do lots of push-ups challenge." Let $(a_n)_{n\geq 1}$ be the sequence that describes the number of push-ups done on the nth day of the challenge. The sequence starts

$$3, 5, 6, 10, 9, 0, 12, \ldots.$$

Describe a sequence $(b_n)_{n\geq 1}$ that describes the total number of push-ups done by Sam after the nth day.

Solution. We can find the terms of this sequence easily enough.

$$3, 8, 14, 24, 33, 33, 45, \ldots.$$

Here b_1 is just a_1, but then

$$b_2 = 3 + 5 = a_1 + a_2,$$

$$b_3 = 3 + 5 + 6 = a_1 + a_2 + a_3,$$

and so on.

There are a few ways we might describe b_n in general. We could do so recursively as,

$$b_n = b_{n-1} + a_n,$$

since the total number of push-ups done after n days will be the number done after $n - 1$ days, plus the number done on day n.

For something closer to a closed formula, we could write

$$b_n = a_1 + a_2 + a_3 + \cdots + a_n,$$

or the same thing using summation notation:

$$b_n = \sum_{i=1}^{n} a_i.$$

However, note that these are not really closed formulas since even if we had a formula for a_n, we would still have an increasing number of computations to do as n increases.

Given any sequence $(a_n)_{n\in\mathbb{N}}$, we can always form a new sequence $(b_n)_{n\in\mathbb{N}}$ by

$$b_n = a_0 + a_1 + a_2 + \cdots + a_n.$$

Since the terms of b_n are the sums of the initial part of the sequence an ways call b_n the **sequence of partial sums of** a_n . Soon we will see that it is sometimes possible to find a closed formula for b_n from the closed formula for a_n .

To simplify writing out these sums, we will often use notation like $\sum_{k=1}^{n} a_k$. This means add up the ak's where k changes from 1 to n.

Example 3.1.6

Use notation to rewrite the sums:

1 + 2 + 3 + 4 +··· + 100

1 + 2 + 4 + 8 +··· + 2^{50}

6 + 10 + 14 +··· + (4n - 2).

Solution

1. $\sum_{k=1}^{100} k$ 2. $\sum_{k=0}^{50} 2^k$ 3. $\sum_{k=2}^{n} (4k-2)$

If we want to multiply the a_k instead, we could write $\prod_{k=1}^{n} a_k$.For example, $\prod_{k=1}^{n} k = n!$.

Exercises

1. **Find the closed formula for each of the following sequences by relating them to a well known sequence. Assume the first term given is a_1.**

   (a) 2, 5, 10, 17, 26, . . . (b) 0, 2, 5, 9, 14, 20, . . .

   (c) , 12, 17, 23, 30, . . . (d) 1, 5, 23, 119, 719, . . .

2. **For each sequence given below, find a closed formula for an, the nth term of the sequence (assume the first terms are a0) by relating it to another sequence for which you already know the formula. In each case, briefly say how you got your answers.**

   (a) 4, 5, 7, 11, 19, 35, . . . (b) 0, 3, 8, 15, 24, 35, . . .

   (c) 6, 12, 20, 30, 42, . . . (d) 0, 2, 7, 15, 26, 40, 57, . . . (Cryptic Hint: these might be called "house numbers")

3. **Write out the first 5 terms (starting with a_0) of each of the sequences described below. Then give either a closed formula or a recursive definition for the sequence (whichever is NOT given in the problem).**

   (a) $a_n = \frac{1}{2}(n^2 + n)$. (b) $a_n = 2a_{n-1} - a_{n-2}$ with $a_0 = 0$ and $a_1 = 1$.

   (c) $a_n = na_{n-1}$ with $a_0 = 1$.

4. **Consider the sequence a_n $(a_n)_{n\geq 1}$ that starts 1, 3, 5, 7, 9, . . . (i.e., the odd numbers in order).**

   (a) Give a recursive definition and closed formula for the sequence.

   (b) Write out the sequence $(b_n)_{n\geq 2}$ of partial sums of (a_n). Write down the recursive definition for (b_n) and guess at the closed formula.

5. **The Fibonacci sequence is 0, 1, 1, 2, 3, 5, 8, 13, . . . (where $F_0 = 0$).**

   (a) Write out the first few terms of the sequence of partial sums: 0, 0 + 1, 0 + 1 + 1,. . .

   (b) Guess a formula for the sequence of partial sums expressed in terms of a single Fibonacci number. For example, you might say $F_0 + F_1 + \cdots + F_n = 3F^2_{n-1} + n$, although that is definitely not correct.

6. **Consider the three sequences below. For each, find a recursive definition. How are these sequences related?**

   (a) 2, 4, 6, 10, 16, 26, 42, ...

   (b) 5, 6, 11, 17, 28, 45, 73, ...

   (c) 0, 0, 0, 0, 0, 0, 0, ...

7. **Write out the first few terms of the sequence given by $a_1 = 3$; $a_n = 2a_{n-1} + 4$. Then find a recursive definition for the sequence 10, 24, 52, 108, ...**

8. **Write out the first few terms of the sequence given by $a_n = n^2\ 3n + 1$. Then find a closed formula for the sequence (starting with a1) 0, 2, 6, 12, 20, ...**

9. **Show that $a_n = 3 \cdot 2n + 7 \cdot 5^n$ is a solution to the recurrence relation $a_n = 7a_{n-1}\ 10a_{n-2}$. What would the initial conditions need to be for this to be the closed formula for the sequence?**

10. **Show that $a_n = 2^n - 5^n$ is also a solution to the recurrence relation $a_n = 7a_n{-1}\ 10a_{n-2}$. What would the initial conditions need to be for this to be the closed formula for the sequence?**

11. **Find a closed formula for the sequence with recursive definition $a_n = 2a_{n-1} - a_{n-2}$ with $a_1 = 1$ and $a_2 = 2$.**

12. **Give two different recursive definitions for the sequence with closed formula $a_n = 3 + 2n$. Prove you are correct. At least one of the recursive definitions should makes use of two previous terms and no constants.**

13. **Use summation $(\sum)$ or product $(\prod)$notation to rewrite the following.**

    (a) $2 + 4 + 6 + 8 + \cdots + 2n$.

    (b) $1 + 5 + 9 + 13 + \cdots + 425$.

    (c) $1 + \frac{1}{2} + \frac{1}{3} + \frac{1}{4} + \cdots + \frac{1}{50}$.

    (d) $2 \cdot 4 \cdot 6\ 2n$.

$(\frac{1}{2})(\frac{2}{3})(\frac{3}{4})\cdots(\frac{100}{101})$.

14. Expand the following sums and products. That is, write them out the long way.

1. $\sum_{k=1}^{100}(3+4k)$.
2. $\sum_{k=0}^{n} 2^k$.
3. $\sum_{k=2}^{50}\frac{1}{(k^2-1)}$.
4. $\prod_{k=2}^{100}\frac{k^2}{(k^2-1)}$.
5. $\prod_{k=0}^{n}(2+3k)$.

15. Suppose you draw n lines in the plane so that every pair of lines cross (no lines are parallel) and no three lines cross at the same point. This will create some number of regions in the plane, including some unbounded regions. Call the number of regions R_n. Find a recursive formula for the number of regions created by n lines, and justify why your recursion is correct.

16. A ternary string is a sequence of 0's, 1's and 2's. Just like a bit string, but with three symbols.

Let's call a ternary string good provided it never contains a 2 followed immediately by a 0. Let G_n be the number of good strings of length n. For example, $G_1 = 3$, and $G_2 = 8$ (since of the 9 ternary strings of length 2, only one is not good).

Find, with justification, a recursive formula for G_n, and use it to compute G_5.

17. Consider bit strings with length l and weight k (so strings of l 0's and 1's, including k 1's). We know how to count the number of these for a fixed l and k. Now, we will count the number of strings for which the *sum* of the length and the weight is fixed. For example, let's count all the bit strings for which l + k = 11.

(a) Find examples of these strings of different lengths. What is the longest string possible? What is the shortest?

(b) How many strings are there of each of these lengths. Use this to count the total number of strings (with sum 11).

(c) The other approach: Let $n = l + k$ vary. How many strings have sum n = 1? How many have sum $n = 2$? And so on. Find and explain a recurrence relation for the sequence (a_n) which gives the number of strings with sum n.

(d) Describe what you have found above in terms of Pascal's Triangle. What pattern have you discovered?

18. **When bees play chess, they use a hexagonal board like the one shown below. The queen bee can move one space at a time either directly to the right or angled up-right or down-right (but can never move leftwards). How many different paths can the queen take from the top left hexagon to the bottom right hexagon? Explain your answer, and this relates to the previous question. (As an example, there are three paths to get to the second hexagon on the bottom row.)**

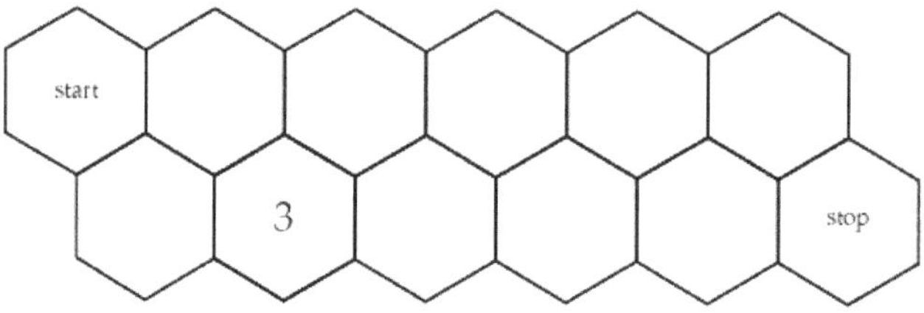

Let t_n denote the number of ways to tile a $2 \times n$ chessboard using 1×2 dominoes. Write out the first few terms of the sequence $(t_n)_{n \geq 1}$ and then give a recursive definition. Explain why your recursive formula is correct.

2.2 ARITHMETIC AND GEOMETRIC SEQUENCES

Investigate!

For the patterns of dots below, draw the next pattern in the sequence. Then give a recursive definition and a closed formula for the number of dots in the nth pattern.

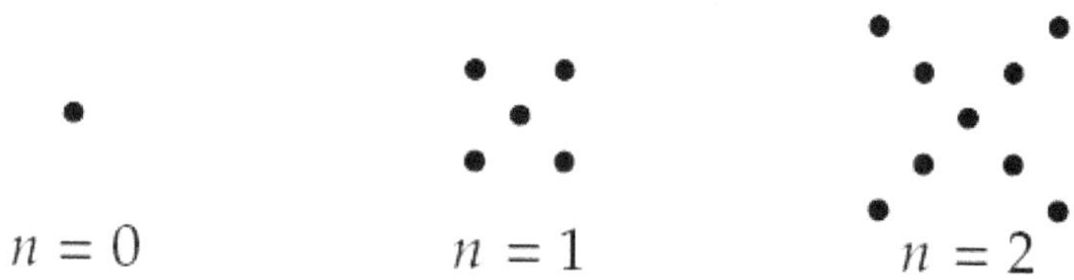

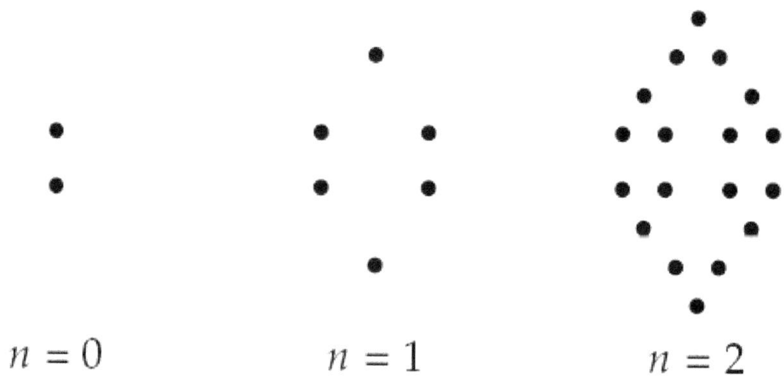

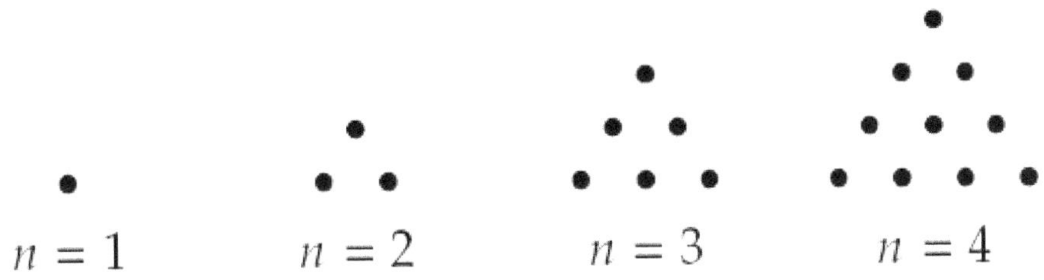

Attempt the above activity before proceeding

We now turn to the question of finding closed formulas for particular types of sequences.

Arithmetic Sequences

If the terms of a sequence differ by a constant, we say the sequence is arithmetic. If the initial term (a_0) of the sequence is a and the **common difference** is d, then we have,

Recursive definition: $a_n = a_{n-1} + d$ with $a_0 = a$.

Closed formula: $a_n = a + dn$.

How do we know this? For the recursive definition, we need to specify a_0. Then we need to express an in terms of a_{n-1}. If we call the first term a, then $a_0 = a$. For the recurrence relation, by the definition of an arithmetic sequence, the difference between successive terms is some constant, say d. So $a_n - a_{n-1} = d$, or in other words,

$$a_0 = a \qquad a_n = a_{n-1} + d.$$

To find a closed formula, first write out the sequence in general:

$$\begin{aligned} a_0 &= a \\ a_1 &= a_0 + d = a + d \\ a_2 &= a_1 + d = a + d + d = a + 2d \\ a_3 &= a_2 + d = a + 2d + d = a + 3d \\ &\vdots \end{aligned}$$

We see that to find the nth term, we need to start with a and then add d a bunch of times. In fact, add it n times. Thus $a_n = a + dn$.

Example 3.2.1

Find recursive definitions and closed formulas for the arithmetic sequences below. Assume the first term listed is a_0.

1. 2, 5, 8, 11, 14, ...
2. 50, 43, 36, 29, ...

Solution

First we should check that these sequences really are arithmetic by taking differences of successive terms. Doing so will reveal the common difference d.

5 - 2 = 3, 8 - 5 = 3, etc. To get from each term to the next, we add three, so d = 3. The recursive definition is therefore $a_n = a_{n-1} + 3$ with $a_0 = 2$. The closed formula is $a_n = 2 + 3n$.

Here the common difference is -7, since we add -7 to 50 to get 43, and so on. Thus we have a recursive definition of $a_n = a_n-1 - 7$ with $a_0 = 50$. The closed formula is $a_n = 50 - 7n$.

What about sequences like 2, 6, 18, 54, . . .? This is not arithmetic because the difference between terms is not constant. However, the ratio between successive terms is constant. We call such sequences **geometric.**

The recursive definition for the geometric sequence with initial term a and common ratio r is $a_n = a_{n-1}\,r$; $a_0 = a$. To get the next term we multiply the previous term by r. We can find the closed formula like we did for the arithmetic progression. Write

$$a_0 = a$$

$$a_1 = a_0 \cdot r$$

$$a_2 = a_1 \cdot r = a_0 \cdot r \cdot r = a_0 \cdot r^2$$

$$\vdots$$

We must multiply the first term a by r a number of times, n times to be precise. We get $a_n = a \cdot r^n$.

Geometric Sequences

A sequence is called **geometric** if the ratio between successive terms is constant. Suppose the initial term a0 is a and the **common ratio** is r. Then we have,

Recursive definition: $a_n = ra_{n-1}$ with $a_0 = a$.

Closed formula: $a_n = a \cdot r^n$.

Example 3.2.2

Find the recursive and closed formula for the geometric sequences below. Again, the first term listed is a_0.

1. 3, 6, 12, 24, 48, . . .
2. 27, 9, 3, 1, 1/3, . . .

Solution

Start by checking that these sequences really are geometric by dividing each term by its previous term. If this ratio really is constant, we will have found *r*.

6/3 = 2, 12/6 = 2, 24/12 = 2, etc. Yes, to get from any term to the next, we multiply by $r = 2$. So the recursive definition is $a_n = 2a_{n-1}$ with $a_0 = 3$. The closed formula is $a_n = 3 \cdot 2^n$.

The common ratio is r = 1/3. So the sequence has recursive definition $a_n = \frac{1}{3}a_{n-1}$ with $a_0=27$ and closed formula $a_n = 27 \cdot \frac{1}{3}^n$.

In the examples and formulas above, we assumed that the initial term was a_0. If your sequence starts with a_1, you can easily find the term that would have been a_0 and use that in the formula. For example, if we want a formula for the sequence 2, 5, 8, . . . and insist that $2 = a_1$, then we can find $a_0 = -1$ (since the sequence is arithmetic with common difference 3, we have $a_0 + 3 = a_1$). Then the closed formula will be $a_n = -1 + 3n$.

Remark 3.2.3 If you look at other textbooks or online, you might find that their closed formulas for arithmetic and geometric sequences differ from ours. Specifically, you might find the formulas $a_n = a + (n-1)d$ (arithmetic) and $a_n = a r^{n-1}$ (geometric). Which is correct? Both! In our case, we take a to be a_0. If instead we had a1 as our initial term, we would get the (slightly more complicated) formulas you find elsewhere.

Sums of Arithmetic and Geometric Sequences

Investigate!

Your neighborhood grocery store has a candy machine full of Skittles.

1 **Suppose that the candy machine currently holds exactly 650 Skittles, and every time someone inserts a quarter, exactly 7 Skittles come out of the machine.**

(a) How many Skittles will be left in the machine after 20 quarters have been inserted?

(b) Will there ever be exactly zero Skittles left in the ma- chine? Explain.

2. **What if the candy machine gives 7 Skittles to the first customer who put in a quarter, 10 to the second, 13 to the third, 16 to the fourth, etc. How many Skittles has the machine given out after 20 quarters are put into the machine?**

3. **Now, what if the machine gives 4 Skittles to the first customer, 7 to the second, 12 to the third, 19 to the fourth, etc. How many Skittles has the machine given out after 20 quarters are put into the machine?**

Attempt the above activity before proceeding

Look at the sequence $(T_n)_{n\geq 1}$ which starts 1, 3, 6, 10, 15, ... These are called the **triangular numbers** since they represent the number of dots in an equilateral triangle (think of how you arrange 10 bowling pins: a row of 4 plus a row of 3 plus a row of 2 and a row of 1).

$T_1 = 1$ $T_2 = 3$ $T_3 = 6$ $T_4 = 10$

Is this sequence arithmetic? No, since 3 1 = 2 and 6 3 = 3 ≠ 2, so there is no common difference. Is the sequence geometric? No. 3 1 = 3 but 6 3 = 2, so there is no common ratio. What to do?

Notice that the *differences* between terms *do* form an arithmetic sequence: 2, 3, 4, 5, 6, ... This means that the nth term of the sequence (T_n) is the *sum* of the first n terms in the sequence 1, 2, 3, 4, 5, ... We say that (T_n) is the **sequence of partial sums** of the sequence 1, 2, 3, (*partial* sums because we are not taking the sum of all infinitely many terms).

This should become clearer if we write the triangular numbers like this:

$$
\begin{aligned}
1 &= 1\\
3 &= 1+2\\
6 &= 1+2+3\\
10 &= 1+2+3+4\\
\vdots\ \ &\ \ \vdots\\
T_n &= 1+2+3+\cdots+n.
\end{aligned}
$$

If we know how to add up the terms of an arithmetic sequence, we could find a closed formula for a sequence whose differences are the terms of that arithmetic sequence. Consider how we could find the sum of the first 100 positive integers (that is, T100). Instead of adding them in order, we regroup and add 1 + 100 = 101. The next pair to combine is 2 + 99 = 101. Then 3 + 98 = 101. Keep going. This gives 50 pairs which each add up to 101, so T_{100} = 101 50 = 5050.[1]

In general, using this same sort of regrouping, we find that $T_n = \frac{n(n+1)}{2}$. Incidentally, this is exactly the same as $\binom{n+1}{2}$,which makes sense if you think of the triangular numbers as counting the number of handshakes that take place at a party with $n + 1$ people: the first person shakes n hands, the next shakes an additional n -1 hands and so on.

The point of all of this is that some sequences, while not arithmetic or geometric, can be interpreted as the sequence of partial sums of arithmetic and geometric sequences. Luckily there are methods we can use to compute these sums quickly.

Summing Arithmetic Sequences: Reverse and Add

Here is a technique that allows us to quickly find the sum of an arithmetic sequence.

Example 3.2.4

Find the sum: $2 + 5 + 8 + 11 + 14 + \cdots + 470$.

Solution. The idea is to mimic how we found the formula for triangular numbers. If we add the first and last terms, we get 472. The second term and second-to-last term also add up to 472. To keep track of everything, we might express this as follows. Call the sum S. Then,

$$\begin{array}{rcccccccccc} & S = & 2 & + & 5 & + & 8 & +\cdots+ & 467 & + & 470 \\ + & S = & 470 & + & 467 & + & 464 & +\cdots+ & 5 & + & 2 \\ \hline & 2S = & 472 & + & 472 & + & 472 & +\cdots+ & 472 & + & 472 \end{array}$$

To find 2S then we add 472 to itself a number of times. What number? We need to decide how many terms (**summands**) are in the sum. Since the terms form an arithmetic sequence, the nth term in the sum (counting 2 as the 0th term) can be expressed as $2 + 3n$.

If $2 + 3n = 470$ then $n = 156$. So n ranges from 0 to 156, giving 157 terms in the sum. This is the number of 472's in the sum for 2S.

Thus

$$2S = 157 \cdot 472 = 74104.$$

It is now easy to find S:

$$S = 74104/2 = 37052.$$

This will work for the sum of any *arithmetic* sequence. Call the sum S. Reverse and add. This produces a single number added to itself many times. Find the number of times. Multiply. Divide by 2. Done.

Example 3.2.5

Find a closed formula for $6 + 10 + 14 + \cdots + (4n - 2)$.

Solution

Again, we have a sum of an arithmetic sequence. How many terms are in the sequence? Clearly each term in the sequence has the form 4k - 2 (as evidenced by the last term). For which values of k though? To get 6, $k = 2$. To get 4n - 2 take $k = n$. So to find the number of terms, we must count the number of integers in the range 2, 3, . . . , n. This is n - 1. (There are n numbers from 1 to n, so one less if we start with 2.)

Now reverse and add:

$$\begin{array}{rrcrcrcrcr} & S = & 6 & + & 10 & +\cdots+ & 4n-6 & + & 4n-2 \\ + & S = & 4n-2 & + & 4n-6 & +\cdots+ & 10 & + & 6 \\ \hline & 2S = & 4n+4 & + & 4n+4 & +\cdots+ & 4n+4 & + & 4n+4 \end{array}$$

Since there are n - 1 terms, we get

$$2S = (n-1)(4n+4) \qquad \text{so} \qquad S = \frac{(n-1)(4n+4)}{2}.$$

Besides finding sums, we can use this technique to find closed formulas for sequences we recognize as sequences of partial sums.

Example 3.2.6

Use partial sums to find a closed formula for $(a_n)_{n\geq 0}$ which starts 2, 3, 7, 14, 24, 37, ...

Solution

First, if you look at the differences between terms, you get a sequence of differences: 1, 4, 7, 10, 13, , which is an arithmetic sequence. Written another way:

$$a_0 = 2$$
$$a_1 = 2 + 1$$
$$a_2 = 2 + 1 + 4$$
$$a_3 = 2 + 1 + 4 + 7$$

and so on. We can write the general term of (a_n) in terms of the arithmetic sequence as follows:

$$a_n = 2 + 1 + 4 + 7 + 10 + \cdots + (1 + 3(n-1))$$

(we use 1 + 3 (n- 1) instead of 1 + 3n to get the indices to line up correctly; for a_3 we add up to 7, which is 1 + 3 (3- 1)).

We can reverse and add, but the initial 2 does not fit our pattern.

This just means we need to keep the 2 out of the reverse part:

$$\begin{array}{rcccccccccc} & a_n = & 2 & + & 1 & + & 4 & + \cdots + & 1+3(n-1) \\ + & a_n = & 2 & + & 1+3(n-1) & + & 1+3(n-2) & + \cdots + & 1 \\ \hline & 2a_n = & 4 & + & 2+3(n-1) & + & 2+3(n-1) & + \cdots + & 2+3(n-1) \end{array}$$

Not counting the first term (the 4) there are n summands of 2 + 3 (*n*- 1) = 3*n* 1 so the right-hand side becomes 4 + (3*n*- 1) n.

Finally, solving for a_n we get

$$a_n = \frac{4 + (3n-1)n}{2}.$$

Just to be sure, we check $a_0 = \frac{4}{2} = 2, a_1 = \frac{4+2}{2} = 3$,etc. We have the correct closed formula.

Summing Geometric Sequences: Multiply, Shift and Subtract

To find the sum of a geometric sequence, we cannot just reverse and add. Do you see why? The reason we got the same term added to itself many times is because there was a constant difference. So as we added that difference in one direction, we subtracted the difference going the other way, leaving a constant total. For geometric sums, we have a different technique.

Example 3.2.7

What is 3 + 6 + 12 + 24 + · · · + 12288?

Solution. Multiply each term by 2, the common ratio. You get 2*S* =6 + 12 + 24 + · · · + 24576. Now subtract: 2*S* - S = -3 + 24576 = 24573.

Since 2*S* - *S* = *S*, we have our answer.

To better see what happened in the above example, try writing it this way:

$$\begin{array}{rrlll} & S = & 3+ & 6+12+24+\cdots+12288 & \\ - & 2S = & & 6+12+24+\cdots+12288 & +24576 \\ \hline & -S = & 3+ & 0+0+0+\cdots+0 & -24576 \end{array}$$

Then divide both sides by -1 and we have the same result for S. The idea is, by multiplying the sum by the common ratio, each term becomes the next term. We shift over the sum to get the subtraction to mostly cancel out, leaving just the first term and new last term.

Example 3.2.8

Find a closed formula for $S(n) = 2 + 10 + 50 + \cdots + 2 \cdot 5^n$.

Solution.

The common ratio is 5. So we have

$$\begin{array}{rrl} & S & = 2 + 10 + 50 + \cdots + 2 \cdot 5^n \\ - & 5S & = \quad 10 + 50 + \cdots + 2 \cdot 5^n + 2 \cdot 5^{n+1} \\ \hline & -4S & = 2 - 2 \cdot 5^{n+1} \end{array}$$

Thus $S = \dfrac{2 - 2 \cdot 5^{n+1}}{-4}$

Even though this might seem like a new technique, you have probably used it before.

Example 3.2.9

Express 0.464646 . . . as a fraction.

Solution.

Let $N = 0.46464646$ Consider $0.01N$. We get:

$$\begin{array}{rrl} & N = & 0.4646464\ldots \\ - & 0.01N = & 0.00464646\ldots \\ \hline & 0.99N = & 0.46 \end{array}$$

So $N = \frac{46}{99}$. What have we done? We viewed the repeating decimal 0.464646 . . . as a sum of the geometric sequence 0.46, 0.0046, 0.000046, . . .

The common ratio is 0.01. The only real difference is that we are now computing an *infinite* geometric sum, we do not have the extra "last" term to consider. Really, this is the result of taking a limit as you would in calculus when you compute *infinite* geometric sums.

Exercises

1. **Consider the sequence 5, 9, 13, 17, 21, . . . with $a_1 = 5$**
   (a) Give a recursive definition for the sequence.
   (b) Give a closed formula for the *n*th term of the sequence.
   (c) Is 2013 a term in the sequence? Explain.
   (d) How many terms does the sequence 5, 9, 13, 17, 21, . . . , 533 have?
   (e) Find the sum: $5 + 9 + 13 + 17 + 21 + \cdots + 533$. Show your work.
   (f) Use what you found above to find bn, the nth term of 1, 6, 15, 28, 45, . . ., where $b_0 = 1$
2. **Consider the sequence (an)n≥0 which starts 8, 14, 20, 26, ...**
   (a) What is the next term in the sequence?

(b) Find a formula for the nth term of this sequence.

(c) Find the sum of the first 100 terms of the sequence: $\sum_{k=0}^{99} a_k$.

3. Consider the sum 4 + 11 + 18 + 25 + · · · + 249.

(a) How many terms (summands) are in the sum?

(b) Compute the sum using a technique discussed in this section.

4. Consider the sequence 1, 7, 13, 19, . . . , 6n + 7.

(a) How many terms are there in the sequence? Your answer will be in terms of n.

(b) What is the second-to-last term?

(c) Find the sum of all the terms in the sequence, in terms of n.

5. Find 5 + 7 + 9 + 11 + · · · + 521 using a technique from this section.

6. Find $5 + 15 + 45 + \cdots + 5 \cdot 3^{20}$.

7. Find $1 - \frac{2}{3} + \frac{4}{9} - \cdots + \frac{2^{30}}{3^{30}}$.

8. Find x and y such that 27, x, y, 1 is part of an arithmetic sequence.

Then find x and y so that the sequence is part of a geometric sequence.

(Warning: x and y might not be integers.)

9. Find x and y such that 5, x, y, 32 is part of an arithmetic sequence.

Then find *x* and *y* so that the sequence is part of a geometric sequence.

(Warning: *x* and *y* might not be integers.)

10. Is there a pair of integers (a, b) such that a, x_1, y_1, b is part of an arithmetic sequences and a, x_2, y_2, b is part of a geometric sequence with x_1, x_2, y_1, y_2 all integers?

11. Consider the sequence 2, 7, 15, 26, 40, 57, . . . (with $a_0 = 2$). By looking at the differences between terms, express the sequence as a sequence of partial sums. Then find a closed formula for the sequence by computing the *n*th partial sum.

12. Starting with any rectangle, we can create a new, larger rectangle by attaching a square to the longer side. For example, if we start with a 2 × 5 rectangle, we would glue on a 5 × 5 square, forming a 5 × 7 rectangle:

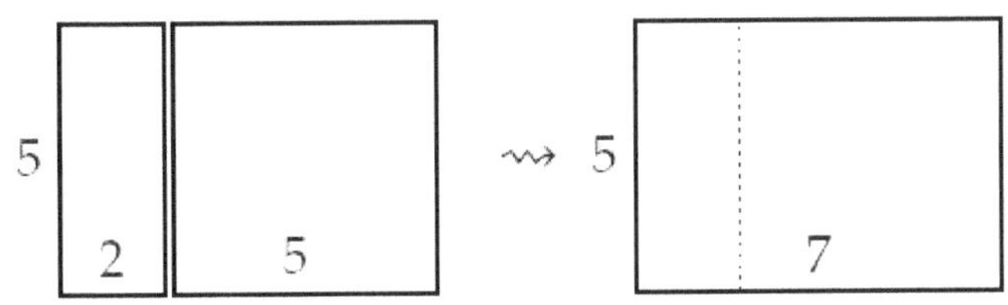

The next rectangle would be formed by attaching a 7 × 7 square to the top or bottom of the 5 × 7 rectangle.

(a) Create a sequence of rectangles using this rule starting with a 1 × 2 rectangle. Then write out the sequence of perimeters for the rectangles (the first term of the sequence would be 6, since the perimeter of a 1 × 2 rectangle is 6 - the next term would be 10).

(b) Repeat the above part this time starting with a 1 × 3 rectangle.

(c) Find recursive formulas for each of the sequences of perimeters you found in parts (a) and (b). Don't forget to give the initial conditions as well.

d) Are the sequences arithmetic? Geometric? If not, are they close to being either of these (i.e., are the differences or ratios almost constant)? Explain.

13. If you have enough toothpicks, you can make a large triangular grid. Below, are the triangular grids of size 1 and of size 2. The size 1 grid requires 3 toothpicks, the size 2 grid requires 9 toothpicks.

(a) Let t_n be the number of toothpicks required to make a size n triangular grid. Write out the first 5 terms of the sequence $t_1, t_2, \ldots$.

(b) Find a recursive definition for the sequence. Explain why you are correct.

(c) Is the sequence arithmetic or geometric? If not, is it the sequence of partial sums of an arithmetic or geometric sequence? Explain why your answer is correct.

(d) Use your results from part (c) to find a closed formula for the sequence. Show your work.

14. If you were to shade in a $n \times n$ square on graph paper, you could do it the boring way (with sides parallel to the edge of the paper) or the interesting way, as illustrated below:

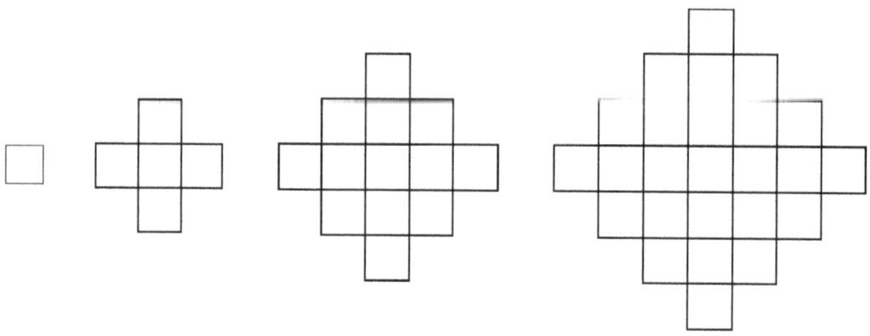

The interesting thing here, is that a 3 × 3 square now has area 13. Our goal is the find a formula for the area of a $n \times n$ (diagonal) square.

(a) Write out the first few terms of the sequence of areas (assume $a_1 = 1$, $a_2 = 5$, etc). Is the sequence arithmetic or geometric? If not, is it the sequence of partial sums of an arithmetic or geometric sequence? Explain why your answer is correct, referring to the diagonal squares.

(b) Use your results from part (a) to find a closed formula for the sequence. Show your work. Note, while there are lots of ways to find a closed formula here, you should use partial sums specifically.

(c) Find the closed formula in as many other interesting ways as you can.

15. **Here is a surprising use of sequences to answer a counting question: How many license plates consist of 6 symbols, using only the three numerals 1, 2, and 3 and the four letters *a*, *b*, *c*, and *d*, so that no numeral appears after any letter? For example, "31ddac", "123321", and "ababab" are each acceptable license plates, but "13ba2c" is not.**

(a) First answer this question by considering different cases: how many of the license plates contain no numerals? How many contain one numeral, etc.

(b) Now use the techniques of this section to show why the answer is $4^7 - 3^7$.

3.3 POLYNOMIAL FITTING

Investigate!

A standard 8 × 8 chessboard contains 64 squares. Actually, this is just the number of unit squares. How many squares of all sizes are there on a chessboard? Start with smaller boards: 1 × 1, 2 × 2, 3 × 3, etc. Find a formula for the total number of squares in an $n \times n$ board.

Attempt the above activity before proceeding

So far we have seen methods for finding the closed formulas for arithmetic and geometric sequences. Since we know how to compute the sum of the first n terms of arithmetic and geometric sequences, we can compute the closed formulas for sequences which have an arithmetic (or geometric) sequence of differences between terms. But what if we consider a sequence which is the sum of the first n terms of a sequence which is itself the sum of an arithmetic sequence?

Before we get too carried away, let's consider an example: How many squares (of all sizes) are there on a chessboard? A chessboard consists of 64 squares, but we also want to consider squares of longer side length. Even though we are only considering an 8 × 8 board, there is already a lot to count. So instead, let us build a sequence: the first term will be the number of squares on a 1 × 1 board, the second term will be the number of squares on a 2 × 2 board, and so on. After a little thought, we arrive at the sequence

$$1, 5, 14, 30, 55, \ldots.$$

This sequence is not arithmetic (or geometric for that matter), but perhaps it's sequence of differences is. For differences we get

$$4, 9, 16, 25, \ldots.$$

Not a huge surprise: one way to count the number of squares in a 4×4 chessboard is to notice that there are 16 squares with side length 1, 9 with side length 2, 4 with side length 3 and 1 with side length 4. So the original sequence is just the sum of squares. Now this sequence of differences is not arithmetic since it's sequence of differences (the differences of the differences of the original sequence) is not constant. In fact, this sequence of **second differences** is which *is* an arithmetic sequence (with constant difference 2). Notice that our original sequence had **third differences** (that is, differences of differences of differences of the original) constant. We will call such a sequence Δ^3-constant. The sequence 1, 4, 9, 16, ... has second differences constant, so it will be a Δ^2-constant sequence. In general, we will say a sequence is a Δk-constant sequence if the kth differences are constant.

$$5, 7, 9, \ldots,$$

Example 3.3.1

Which of the following sequences are Δ^k-constant for some value of k?

1. 2, 3, 7, 14, 24, 37, ...
2. 1, 8, 27, 64, 125, 216, ...
3. 1, 2, 4, 8, 16, 32, 64, ...

Solution

This is the sequence from Example 2.2.6, in which we found a closed formula by recognizing the sequence as the sequence of partial sums of an arithmetic sequence. Indeed, the sequence of first differences is 1, 4, 7, 10, 13, . . ., which itself has differences 3, 3, 3, 3, ... Thus 2, 3, 7, 14, 24, 37, . . . is a Δ^2-constant sequence.

These are the perfect cubes. The sequence of first differences is 7, 19, 37, 61, 91, . . .; the sequence of second differences is 12, 18, 24, 30, . . .; the sequence of third differences is constant: 6, 6, 6, Thus the perfect cubes are a Δ^3-constant sequence.

If we take first differences we get 1, 2, 4, 8, 16, Wait, what? That's the sequence we started with. So taking second differences will give us the same sequence again. No matter how many times we repeat this we will always have the same sequence, which in particular means no finite number of differences will be constant. Thus this sequence is not Δk-constant for any k.

The Δ^0-constant sequences are themselves constant, so a closed formula for them is easy to compute (it's just the constant). The Δ^1-constant sequences are arithmetic and we have a method for finding closed formulas for them as well. Every Δ^2-constant sequence is the sum of an arithmetic sequence so we can find formulas for these as well. But notice that the format of the closed formula for a Δ^2-constant sequence is always quadratic. For example, the square numbers are Δ^2-constant with closed formula $a_n = n^2$. The triangular numbers (also Δ^2-constant) have closed formula $a_n = \frac{n(n+1)}{2}$, which when multiplied out gives you an n^2 term as well. It appears that every time we increase the complexity of the sequence, that is, increase the number of differences before we get constants, we also increase the degree of the polynomial used for the closed formula. We go from constant to linear to quadratic. The sequence of differences between terms tells us something about the rate of growth of the sequence. If a sequence is growing at a constant rate, then the formula for the sequence will be linear. If the sequence is growing at a rate which itself is growing at a constant rate, then the formula is quadratic. You have seen this elsewhere: if a function has a constant second derivative (rate of change) then the function must be quadratic.

This works in general:

Finite Differences

The closed formula for a sequence will be a degree k polynomial if and only if the sequence is Δ^k-constant (i.e., the kth sequence of differences is constant).

This tells us that the sequence of numbers of squares on a chessboard, 1, 5, 14, 30, 55, . . ., which we saw to be Δ^3-constant, will have a cubic (degree 3 polynomial) for its closed formula.

Now once we know what format the closed formula for a sequence will take, it is much easier to actually find the closed formula. In the case that the closed formula is a degree k polynomial, we just need k + 1 data points to "fit" the polynomial to the data.

Example 3.3.2

Find a formula for the sequence 3, 7, 14, 24, ... Assume $a_1 = 3$.

Solution

First, check to see if the formula has constant differences at some level. The sequence of first differences is 4, 7, 10, ... which is arithmetic, so the sequence of second differences is constant. The sequence is Δ^2-constant, so the formula for an will be a degree 2 polynomial. That is, we know that for some constants a, b, and c,

$$a_n = an^2 + bn + c.$$

Now to find a, b, and c. First, it would be nice to know what a_0 is, since plugging in n = 0 simplifies the above formula greatly. In this case, a_0 = 2 (work backwards from the sequence of constant differences). Thus

$$a_0 = 2 = a \cdot 0^2 + b \cdot 0 + c,$$

so $c = 2$. Now plug in $n = 1$ and $n = 2$. We get

$$a_1 = 3 = a + b + 2$$

$$a_2 = 7 = a4 + b2 + 2.$$

At this point we have two (linear) equations and two unknowns, so we can solve the system for a and b (using substitution or elimination or even matrices). We find $a = \frac{3}{2}$ and $b = \frac{-1}{2}$, so

$$a_n = \frac{3}{2}n^2 - \frac{1}{2}n + 2.$$

Example 3.3.3

Find a closed formula for the number of squares on an n × n chessboard.

Solution

We have seen that the sequence 1, 5, 14, 30, 55, . . . is Δ^3-constant, so we are looking for a degree 3 polynomial. That is,

$$a_n = an^3 + bn^2 + cn + d.$$

We can find d if we know what a0 is. Working backwards from the third differences, we find $a_0 = 0$ (unsurprisingly, since there are no squares on a 0 × 0 chessboard). Thus d = 0. Now plug in $n = 1$, $n = 2$, and $n = 3$:

$$1 = a + b + c$$
$$5 = 8a + 4b + 2c$$
$$14 = 27a + 9b + 3c.$$

If we solve this system of equations we get $a = \frac{1}{3}$, $b = \frac{1}{2}$ and $c = \frac{1}{6}$. Therefore the number of squares on an $n \times n$ chessboard is

$$a_n = \frac{1}{3}n^3 + \frac{1}{2}n^2 + \frac{1}{6}n = \frac{1}{6}n(n + 1)(2n + 1).$$

Note: Since the squares-on-a-chessboard problem is really asking for the sum of squares, we now have a nice formula for $\sum_{k=1}^{n} k^2$.

Not all sequences will have polynomials as their closed formula. We can use the theory of finite differences to identify these.

Example 3.3.4

Determine whether the following sequences can be described by a polynomial, and if so, of what degree.

$$1, 2, 4, 8, 16, \ldots$$

$$0, 7, 50, 183, 484, 1055, \ldots$$

$$1, 1, 2, 3, 5, 8, 13, \ldots$$

Solution

As we saw in Example 2.3.1, this sequence is not Δ^k-constant for any k. Therefore the closed formula for the sequence is not a polynomial. In fact, we know the closed formula is $a_n = 2^n$, which grows faster than any polynomial (so is not a polynomial).

The sequence of first differences is 7, 43, 133, 301, 571, ... The second differences are: 36, 90, 168, 270, Third difference: 54, 78, 102,. . .. Fourth differences: 24, 24, As far as we can tell, this sequence of differences is constant so the sequence is Δ^4-constant and as such the closed formula is a degree 4 polynomial.

This is the Fibonacci sequence. The sequence of first differences is 0, 1, 1, 2, 3, 5, 8, . . ., the second differences are 1, 0, 1, 1, 2, 3, 5 . . .. We notice that after the first few terms, we get the original sequence back. So there will never be constant differences, so the closed formula for the Fibonacci sequence is not a polynomial.

Exercises

1. **Use polynomial fitting to find the formula for the nth term of the sequence $(a_n)_{n\geq 0}$ which starts,**

$$0, 2, 6, 12, 20, \ldots.$$

2. **Use polynomial fitting to find the formula for the nth term of the sequence $(a_n)_{n\geq 0}$ which starts,**

$$1, 2, 4, 8, 15, 26 \ldots.$$

3. **Use polynomial fitting to find the formula for the nth term of the sequence $(a_n)_{n\geq 0}$ which starts,**

   2, 5, 11, 21, 36, ...

4. **Use polynomial fitting to find the formula for the nth term of the sequence $(a_n)_{n\geq 0}$ which starts,**

$$3, 6, 12, 22, 37, 58, \ldots.$$

5. **Make up sequences that have**
   (a) 3, 3, 3, 3, . . . as its second differences.
   (b) 1, 2, 3, 4, 5, . . . as its third differences.
   (c) 1, 2, 4, 8, 16, . . . as its 100th differences.

6. **Consider the sequence 1, 3, 7, 13, 21, ... Explain how you know the closed formula for the sequence will be quadratic. Then "guess" the correct formula by comparing this sequence to the squares 1, 4, 9, 16, . . . (do not use polynomial fitting).**

7. **Use a similar technique as in the previous exercise to find a closed formula for the sequence 2, 11, 34, 77, 146, 247, ...**
8. **Suppose $a_n = n^2 + 3n + 4$. Find a closed formula for the sequence of differences by computing $a_n - a_{n-1}$.**
9. **Generalize Exercise 2.3.8: Find a closed formula for the sequence of differences of $a_n = an^2 + bn + c$. That is, prove that every quadratic sequence has arithmetic differences.**
10. **Can you use polynomial fitting to find the formula for the nth term of the sequence 4, 7, 11, 18, 29, 47,...... ? Explain why or why not.**
11. **Will the nth sequence of differences of 2, 6, 18, 54, 162, . . . ever be constant? Explain.**
12. **In their down time, ghost pirates enjoy stacking cannonballs in trian- gular based pyramids (aka, tetrahedrons), like those pictured here:**

Note, these are solid tetrahedrons, so there will be some cannonballs obscured from view (the picture on the right has one cannonball in the back not shown in the picture, for example)

The pirates wonder how many cannonballs would be required to build a pyramid 15 layers high (thus breaking the world cannonball stacking record). Can you help?

(a) Let $P(n)$ denote the number of cannonballs needed to create a pyramid n layers high. So $P(1) = 1$, $P(2) = 4$, and so on. Calculate $P(3)$, $P(4)$ and $P(5)$.

(b) Use polynomial fitting to find a closed formula for $P(n)$. Show your work.

(c) Answer the pirate's question: how many cannonballs do they need to make a pyramid 15 layers high?

(d) Bonus: Locate this sequence in Pascal's triangle. Why does that make sense?

3.4 SOLVING RECURRENCE RELATIONS

Investigate!

Consider the recurrence relation

$$a_n = 5a_{n-1} - 6a_{n-2}.$$

- What sequence do you get if the initial conditions are $a_0 = 1$, $a_1 = 2$? Give a closed formula for this sequence.
- What sequence do you get if the initial conditions are $a_0 = 1$, $a_1 = 3$? Give a closed formula.
- What if $a_0 = 2$ and $a_1 = 5$? Find a closed formula.

Attempt the above activity before proceeding

We have seen that it is often easier to find recursive definitions than closed formulas. Lucky for us, there are a few techniques for converting recursive definitions to closed formulas. Doing so is called **solving a recurrence relation.** Recall that the recurrence relation is a recursive definition without the initial conditions. For example, the recurrence relation for the Fibonacci sequence is $F_n = F_{n-1} + F_{n-2}$. (This, together with the initial conditions $F_0 = 0$ and $F_1 = 1$ give the entire recursive *definition* for the sequence.)

Example 3.4.1

Find a recurrence relation and initial conditions for 1, 5, 17, 53, 161, 485 . . ..

Solution

Finding the recurrence relation would be easier if we had some context for the problem (like the Tower of Hanoi, for example). Alas, we have only the sequence. Remember, the recurrence relation tells you how to get from previous terms to future terms. What is going on here? We could look at the differences between terms: 4, 12, 36, 108, ... Notice that these are growing by a factor of 3. Is the original sequence as well? $1 \cdot 3 = 3$, $5 \cdot 3 = 15$, $17 \cdot 3 = 51$ and so on. It appears that we always end up with 2 less than the next term. Aha!

So $a_n = 3a_{n-1} + 2$ is our recurrence relation and the initial condition is $a_0 = 1$.

We are going to try to solve these recurrence relations. By this we mean something very similar to solving differential equations: we want to find a function of n (a closed formula) which satisfies the recurrence relation, as well as the initial condition.[2] Just like for differential equations, finding a solution might be tricky, but checking that the solution is correct is easy.

Example 3.4.2

Check that $a_n = 2^n + 1$ is a solution to the recurrence relation $a_n = 2a_{n-1} - 1$ with $a_1 = 3$.

Solution

First, it is easy to check the initial condition: a1 should be $2^1 + 1$ according to our closed formula. Indeed, 21 + 1 = 3, which is what we want. To check that our proposed solution satisfies the recurrence relation, try plugging it in.

$$\begin{aligned} 2a_{n-1} - 1 &= 2(2^{n-1} + 1) - 1 \\ &= 2^n + 2 - 1 \\ &= 2^n + 1 \\ &= a_n. \end{aligned}$$

That's what our recurrence relation says! We have a solution.

Sometimes we can be clever and solve a recurrence relation by inspection. We generate the sequence using the recurrence relation and keep track of what we are doing so that we can see how to jump to finding just the a_n term. Here are two examples of how you might do that.

Telescoping refers to the phenomenon when many terms in a large sum cancel out—so the sum "telescopes." For example:

$$(2-1) + (3-2) + (4-3) + \cdots + (100-99) + (101-100) = -1 + 101$$

because every third term looks like: 2 + 2 = 0, and then 3 + – 3 = 0 and so on.

We can use this behavior to solve recurrence relations. Here is an example.

Example 3.4.3

Solve the recurrence relation $a_n = a_{n-1} + n$ with initial term $a_0 = 4$.

Solution

To get a feel for the recurrence relation, write out the first few terms of the sequence: 4, 5, 7, 10, 14, 19, ... Look at the difference between terms. $a_1 - a_0 = 1$ and $a_2 - a_1 = 2$ and so on.

The key thing here is that the difference between terms is n. We can write this explicitly: $a_n - a_{n-1} = n$. Of course, we could have arrived at this conclusion directly from the recurrence relation by subtracting a_{n-1} from both sides.

Now use this equation over and over again, changing n each time:

$$\begin{aligned} a_1 - a_0 &= 1 \\ a_2 - a_1 &= 2 \\ a_3 - a_2 &= 3 \\ \vdots \quad & \quad \vdots \\ a_n - a_{n-1} &= n. \end{aligned}$$

Add all these equations together. On the right-hand side, we get the sum 1 + 2 + 3 + + n. We already know this can be simplified to $\frac{n(n+1)}{2}$. What happens on the left-hand side? We get

$$(a_1 - a_0) + (a_2 - a_1) + (a_3 - a_2) + \cdots (a_{n-1} - a_{n-2}) + (a_n - a_{n-1}).$$

This sum telescopes. We are left with only the a0 from the first equation and the an from the last equation. Putting this all together we have $-a_0 + a_n = \frac{n(n+1)}{2}$ or $a_n = \frac{n(n+1)}{2} + a_0$. But we know that $a_0 = 4$. So the solution to the recurrence relation, subject to the initial condition is

$$a_n = \frac{n(n+1)}{2} + 4.$$

(Now that we know that, we should notice that the sequence is the result of adding 4 to each of the triangular numbers.)

The above example shows a way to solve recurrence relations of the form $a_n = a_{n-1} + f(n)$ where $\sum_{k=1}^{n} f(k)$ has a known closed formula. If you rewrite the recurrence relation as $a_n\ a_{n-1} = f(n)$, and then add up all the different equations with n ranging between 1 and n, the left-hand side will always give you $a_n - a_0$. The right-hand side will be $\sum_{k=1}^{n} f(k)$, which is why we need to know the closed formula for that sum.

However, telescoping will not help us with a recursion such as $a_n = 3a_{n-1} + 2$ since the left-hand side will not telescope. You will have $-3a_{n-1}$'s but only one a_{n-1}. However, we can still be clever if we use **iteration.**

We have already seen an example of iteration when we found the closed formula for arithmetic and geometric sequences. The idea is, we iterate the process of finding the next term, starting with the known initial condition, up until we have an. Then we simplify. In the arithmetic sequence example, we simplified by multiplying d by the number of times we add it to a when we get to a_n, to get from $a_n = a + d + d + d + \dots\dots\dots + d$ to $a_n = a + dn$.

To see how this works, let's go through the same example we used for telescoping, but this time use iteration.

Example 3.4.4

Use iteration to solve the recurrence relation $a_n = a_{n-1} + n$ with $a_0 = 4$.

Solution

Again, start by writing down the recurrence relation when $n = 1$. This time, don't subtract the an−1 terms to the other side:

$$a_1 = a_0 + 1.$$

Now $a_2 = a_1 + 2$, but we know what a_1 is. By substitution, we get

$$a_2 = (a_0 + 1) + 2.$$

Now go to $a_3 = a_2 + 3$, using our known value of a_2:

$$a_3 = ((a_0 + 1) + 2) + 3.$$

We notice a pattern. Each time, we take the previous term and add the current index. So

$$a_n = ((((a_0 + 1) + 2) + 3) + \cdots + n - 1) + n.$$

Regrouping terms, we notice that a_n is just a_0 plus the sum of the integers from 1 to n. So, since $a_0 = 4$,

$$a_n = 4 + \frac{n(n+1)}{2}.$$

Of course in this case we still needed to know formula for the sum of 1, . . . , n. Let's try iteration with a sequence for which telescoping doesn't work.

Example 3.4.5

Solve the recurrence relation $a_n = 3a_{n-1} + 2$ subject to $a_0 = 1$.

Solution

Again, we iterate the recurrence relation, building up to the index n.

$$\begin{aligned}
a_1 &= 3a_0 + 2\\
a_2 &= 3(a_1) + 2 = 3(3a_0 + 2) + 2\\
a_3 &= 3[a_2] + 2 = 3[3(3a_0 + 2) + 2] + 2\\
&\vdots\\
a_n &= 3(a_{n-1}) + 2 = 3(3(3(3\cdots(3a_0 + 2) + 2) + 2)\cdots + 2) + 2.
\end{aligned}$$

It is difficult to see what is happening here because we have to distribute all those 3's. Let's try again, this time simplifying a bit as we go.

$$\begin{aligned}
a_1 &= 3a_0 + 2\\
a_2 &= 3(a_1) + 2 = 3(3a_0 + 2) + 2 = 3^2a_0 + 2\cdot 3 + 2\\
a_3 &= 3[a_2] + 2 = 3[3^2a_0 + 2\cdot 3 + 2] + 2 = 3^3a_0 + 2\cdot 3^2 + 2\cdot 3 + 2\\
&\vdots\\
a_n &= 3(a_{n-1}) + 2 = 3(3^{n-1}a_0 + 2\cdot 3^{n-2} + \cdots + 2) + 2\\
&= 3^na_0 + 2\cdot 3^{n-1} + 2\cdot 3^{n-2} + \cdots + 2\cdot 3 + 2.
\end{aligned}$$

Now we simplify. a0 = 1, so we have 3n + stuff . Note that all the other terms have a 2 in them. In fact, we have a geometric sum with first term 2 and common ratio 3. We have seen how to simplify $2 + 2\cdot 3 + 2\cdot 3^2 + \cdots + 2\cdot 3^{n-1}$. We get $\frac{2-2\cdot 3^n}{-2}$ which simplifies to $3^n - 1$.

Putting this together with the first 3^n term gives our closed formula:

$$a_n = 2\cdot 3^n - 1.$$

Iteration can be messy, but when the recurrence relation only refers to one previous term (and maybe some function of n) it can work well. However, trying to iterate a recurrence relation such as an = $2a_{n-1} + 3a_{n-2}$ will be way too complicated.

We would need to keep track of two sets of previous terms, each of which were expressed by two previous terms, and so on. The length of the formula would grow exponentially (double each time, in fact). Luckily there happens to be a method for solving recurrence relations which works very well on relations like this.

The Characteristic Root Technique

Suppose we want to solve a recurrence relation expressed as a combination of the two previous terms, such as $a_n = a_{n-1} + 6a_{n-2}$. In other words, we want to find a function of n which satisfies $a_n\ a_{n-1}\ 6a_{n-2} = 0$. Now iteration is too complicated, but think just for a second what would happen if we did iterate. In each step, we would, among other things, multiply a previous iteration by 6. So our closed formula would include 6 multiplied some number of times. Thus it is reasonable to guess the solution will contain parts that look geometric. Perhaps the solution will take the form r^n for some constant r.

The nice thing is, we know how to check whether a formula is actually a solution to a recurrence relation: plug it in. What happens if we plug in r^n into the recursion above? We get

$$r^n - r^{n-1} - 6r^{n-2} = 0.$$

Now solve for r:

$$r^{n-2}(r^2 - r - 6) = 0,$$

so by factoring, $r = -2$ or $r = 3$ (or $r = 0$, although this does not help us). This tells us that $a_n = (-2)^n$ is a solution to the recurrence relation, as is $a_n = 3^n$. Which one is correct? They both are, unless we specify initial conditions. Notice we could also have $a_n =(-2)^n + 3^n$. Or $a_n = 7(-2)^n + 4\ 3^n$. In fact, for any a and b, $a_n = a(-2)^n + b3^n$ is a solution (try plugging this into the recurrence relation). To find the values of a and b, use the initial conditions.

This points us in the direction of a more general technique for solving recurrence relations. Notice we will always be able to factor out the r^{n-2} as we did above. So we really only care about the other part. We call this other part the **characteristic equation** for the recurrence relation. We are interested in finding the roots of the characteristic equation, which are called (surprise) the **characteristic roots.**

Characteristic Roots.

Given a recurrence relation $a_n + \alpha a_{n-1} + \beta a_{n-2} = 0$, the **characteristic polynomial** is

$$x^2 + \alpha x + \beta$$

giving the **characteristic equation:**

$$x^2 + \alpha x + \beta = 0.$$

If r^1 and r^2 are two distinct roots of the characteristic polynomial (i.e., solutions to the characteristic equation), e recurrence relation is

$$a_n = ar_1^n + br_2^n,$$

where a and b are constants determined by the initial conditions.

Example 3.4.6

Solve the recurrence relation $a_n = 7a_{n-1}$ $10a_{n-2}$ with $a_0 = 2$ and $a_1 = 3$.

Solution

Rewrite the recurrence relation a_n $7a_{n-1} + 10a_{n-2} = 0$. Now form the characteristic equation:

$$x^2 - 7x + 10 = 0$$

and solve for x:

$(x - 2)(x - 5) = 0$

so $x = 2$ and $x = 5$ are the characteristic roots. We therefore know that the solution to the recurrence relation will have the form

$$a_n = a2^n + b5^n.$$

To find a and b, plug in n = 0 and n = 1 to get a system of two equations with two unknowns:

$$2 = a2^0 + b5^0 = a + b$$

$$3 = a2^1 + b5^1 = 2a + 5b$$

Solving this system gives $a = \frac{7}{3}$ and $b = -\frac{1}{3}$.so the solution to the recurrence relation is

$$a_n = \frac{7}{3}2^n - \frac{1}{3}5^n.$$

Perhaps the most famous recurrence relation is $F_n = F_{n-1} + F_{n-2}$, which together with the initial conditions $F_0 = 0$ and $F_1 = 1$ defines the Fibonacci sequence. But notice that this is precisely the type of recurrence relation on which we can use the characteristic root technique. When you do, the only thing that changes is that the characteristic equation does not factor, so you need to use the quadratic formula to find the characteristic roots.

In fact, doing so gives the third most famous irrational number, ϕ, the **golden ratio.**

Before leaving the characteristic root technique, we should think about what might happen when you solve the characteristic equation. We have an example above in which the characteristic polynomial has two distinct roots. These roots can be integers, or perhaps irrational numbers (requiring the quadratic formula to find them). In these cases, we know what the solution to the recurrence relation looks like.

However, it is possible for the characteristic polynomial to have only one root. This can happen if the characteristic polynomial factors as $(x - r)^2$.

It is still the case that r^n would be a solution to the recurrence relation, but we won't be able to find solutions for all initial conditions using the general form $a_n = ar_1^n + br_2^n$, since we can't distinguish between r^n_1 and r^n_2.

Characteristic Root Technique for Repeated Roots.

Suppose the recurrence relation $a_n = \alpha a_{n-1} + \beta a_{n-2}$ has a characteristic polynomial with only one root r. Then the solution to the recurrence relation is

$$a_n = ar^n + bnr^n$$

where a and b are constants determined by the initial conditions.

Notice the extra n in bnr^n. This allows us to solve for the constants a and b from the initial conditions.

Example 3.4.7

Solve the recurrence relation $a_n = 6a_{n-1}\ 9a_{n-2}$ with initial conditions $a_0 = 1$ and $a_1 = 4$.

Solution.

The characteristic polynomial is $x^2\ 6x + 9$. We solve the characteristic equation

$$x^2 - 6x + 9 = 0$$

by factoring:

$$(x - 3)^2 = 0$$

so $x = 3$ is the only characteristic root. Therefore we know that the solution to the recurrence relation has the form

$$a_n = a3^n + bn3^n$$

for some constants a and b. Now use the initial conditions:

$$a_0 = 1 = a3^0 + b \cdot 0 \cdot 3^0 = a$$
$$a_1 = 4 = a \cdot 3 + b \cdot 1 \cdot 3 = 3a + 3b.$$

Since $a = 1$, we find that $b = \frac{1}{3}$. Therefore the solution to the recurrence relation is

$$a_n = 3^n + \frac{1}{3} n 3^n.$$

Although we will not consider examples more complicated than these, this characteristic root technique can be applied to much more complicated recurrence relations. For example, $a_n = 2a_{n-1} + a_{n-2}\ 3a_{n-3}$ has characteristic polynomial $x^3\ 2x^2\ x + 3$. Assuming you see how to factor such a degree 3 (or more) polynomial you can easily find the characteristic roots and as such solve the recurrence relation (the solution would look like $a_n = ar_1^n + br_2^n + cr_3^n$ if there were 3 distinct roots). It is also possible that the characteristics roots are complex numbers.

However, the characteristic root technique is only useful for solving recurrence relations in a particular form: a_n is given as a linear combination of some number of previous terms. These recurrence relations are called **linear homogeneous recurrence relations with constant coefficients.** The "homogeneous" refers to the fact that there is no additional term in the recurrence relation other than a multiple of a j terms. For example, $a_n = 2a_{n-1} + 1$ is *non-homogeneous* because of the additional constant 1. There are general methods of solving such things, but we will not consider them here, other than through the use of telescoping or iteration described above.

Exercises

1. **Find the next two terms in $(a_n)_{n\geq 0}$ beginning 3, 5, 11, 21, 43, 85 . . ..**

   Then give a recursive definition for the sequence. Finally, use the characteristic root technique to find a closed formula for the sequence.

2. **Consider the sequences 2, 5, 12, 29, 70, 169, 408, (with $a_0 = 2$).**
   (a) Describe the rate of growth of this sequence.
   (b) Find a recursive definition for the sequence.
   (c) Find a closed formula for the sequence.
   (d) If you look at the sequence of differences between terms, and then the sequence of second differences, the sequence of third differences, and so on, will you ever get a constant sequence? Explain how you know.

3. **Solve the recurrence relation $a_n = a_{n-1} + 2n$ with $a_0 = 5$.**

4. **Show that 4n is a solution to the recurrence relation $a_n = 3a_{n-1} + 4a_{n-2}$.**

5. **Find the solution to the recurrence relation $a_n = 3a_{n-1} + 4a_{n-2}$ with initial terms $a_0 = 2$ and $a_1 = 3$.**

6. **Find the solution to the recurrence relation $a_n = 3a_{n-1} + 4a_{n-2}$ with initial terms $a_0 = 5$ and $a_1 = 8$.**

7. **Solve the recurrence relation $a_n = 3a_{n-1} + 10a_{n-2}$ with initial terms $a_0 = 4$ and $a_1 = 1$.**
8. **Suppose that r^n and q^n are both solutions to a recurrence relation of the form $a_n = \alpha a_{n-1} + \beta a_{n-2}$. Prove that $c. r^n + d. q^n$ is also a solution to the recurrence relation, for any constants c, d.**
9. **Think back to the magical candy machine at your neighborhood grocery store. Suppose that the first time a quarter is put into the machine 1 Skittle comes out. The second time, 4 Skittles, the third time 16 Skittles, the fourth time 64 Skittles, etc.**
   (a) Find both a recursive and closed formula for how many Skittles the nth customer gets.
   (b) Check your solution for the closed formula by solving the recur- rence relation using the Characteristic Root technique.
10. **Let a_n be the number of $1 \times n$ tile designs you can make using 1×1 squares available in 4 colors and 1×2 dominoes available in 5 colors.**
    (a) First, find a recurrence relation to describe the problem. Explain why the recurrence relation is correct (in the context of the problem).
    (b) Write out the first 6 terms of the sequence a1, a2, ...
    (c) Solve the recurrence relation. That is, find a closed formula.
11. **You have access to 1×1 tiles which come in 2 different colors and 1×2 tiles which come in 3 different colors. We want to figure out how many different $1 \times n$ path designs we can make out of these tiles.**
    (a) Find a recursive definition for the number of paths of length n.
    (b) Solve the recurrence relation using the Characteristic Root technique.
12. **Solve the recurrence relation $a_n = 2a_{n-1} - a_{n-2}$.**
    (a) What is the solution if the initial terms are $a_0 = 1$ and $a_1 = 2$?
    (b) What do the initial terms need to be in order for $a_9 = 30$?
    (c) For which x are there initial terms which make $a_9 = x$?
13. **Consider the recurrence relation $a_n = 4a_{n-1} - 4a_{n-2}$.**
    (a) Find the general solution to the recurrence relation (beware the repeated root).
    (b) Find the solution when $a_0 = 1$ and $a_1 = 2$.
    (c) Find the solution when $a_0 = 1$ and $a_1 = 8$.

3.5 INDUCTION

Mathematical induction is a proof technique, not unlike direct proof or proof by contradiction or combinatorial proof.[3] In other words, induction is a style of argument we use to convince ourselves and others that a mathematical statement is always true. Many mathematical statements can be proved by simply explaining what they mean. Others are very difficult to prove—in fact, there are relatively simple mathematical statements which nobody yet knows how to prove. To facilitate the discovery of proofs, it is important to be familiar with some standard styles of arguments. Induction is one such style. Let's start with an example:

Stamps

Investigate!

You need to mail a package, but don't yet know how much postage you will need. You have a large supply of 8-cent stamps and 5-cent stamps. Which amounts of postage can you make exactly using these stamps? Which amounts are impossible to make?

Attempt the above activity before proceeding

Perhaps in investigating the problem above you picked some amounts of postage, and then figured out whether you could make that amount using just 8-cent and 5-cent stamps. Perhaps you did this in order: can you make 1 cent of postage? Can you make 2 cents? 3 cents? And so on. If this is what you did, you were actually answering a sequence of questions. We have methods for dealing with sequences. Let's see if that helps.

Actually, we will not make a sequence of questions, but rather a sequence of statements. Let P (n) be the statement "you can make n cents of postage using just 8-cent and 5-cent stamps." Since for each value of n, $P(n)$ is a statement, it is either true or false. So if we form the sequence of statements

$$P(1), P(2), P(3), P(4), \ldots,$$

the sequence will consist of T's (for true) and F's (for false). In our particular case the sequence starts

$$F, F, F, F, T, F, F, T, F, T, F, F, T, \ldots$$

because $P(1)$, $P(2)$, $P(3)$, $P(4)$ are all false (you cannot make 1, 2, 3, or 4 cents of postage) but $P(5)$ is true (use one 5-cent stamp), and so on.

Let's think a bit about how we could find the value of $P(n)$ for some specific n (the "value" will be either T or F). How did we find the value of the nth term of a sequence of numbers? How did we find a_n? There were two ways we could do this: either there was a closed formula for a_n, so we could plug in n into the formula and get our output value, or we had a recursive definition for the sequence, so we could use the previous terms of the sequence to compute the nth term. When dealing with sequences of statements, we could use either of these techniques as well. Maybe there is a way to use n itself to determine whether we can make n cents of postage. That would be something like a closed formula. Or instead we could use the previous terms in the sequence (of statements) to determine whether we can make n cents of postage. That is, if we know the value of $P(n-1)$, can we get from that to the value of $P(n)$? That would be something like a recursive definition for the sequence. Remember, finding recursive definitions for sequences was often easier than finding closed formulas. The same is true here.

Suppose I told you that $P(43)$ was true (it is). Can you determine from this fact the value of $P(44)$ (whether it true or false)? Yes you can. Even if we don't know how exactly we made 43 cents out of the 5-cent and 8-cent stamps, we do know that there was some way to do it. What if that way used at least three 5-cent stamps (making 15 cents)? We could replace those three 5-cent stamps with two 8-cent stamps (making 16 cents). The total postage has gone up by 1, so we have a way to make 44 cents, so $P(44)$ is true. Of course, we assumed that we had at least three 5-cent stamps. What if we didn't? Then we must have at least three 8-cent stamps (making 24 cents). If we replace those three 8-cent stamps with five 5-cent stamps (making 25 cents) then again we have bumped up our total by 1 cent so we can make 44 cents, so $P(44)$ is true.

Notice that we have not said how to make 44 cents, just that we can, on the basis that we can make 43 cents. How do we know we can make 43 cents? Perhaps because we know we can make 42 cents, which we know we can do because we know we can make 41 cents, and so on. It's a recursion! As with a recursive definition of a numerical sequence, we must specify our initial value. In this case, the initial value is "$P(1)$ is false."

That's not good, since our recurrence relation just says that $P(k+1)$ is true *if* $P(k)$ is also true. We need to start the process with a true $P(k)$. So instead, we might want to use "$P(28)$ is true" as the initial condition.

Putting this all together we arrive at the following fact: it is possible to (exactly) make any amount of postage greater than 27 cents using just 5-cent and 8-cent stamps.[4] In other words, $P(k)$ is true for any $k \geq 28$. To prove this, we could do the following:

- Demonstrate that $P(28)$ is true.
- Prove that if $P(k)$ is true, then $P(k+1)$ is true (for any $k \geq 28$).

Suppose we have done this. Then we know that the 28th term of the sequence above is a T (using step 1, the initial condition or **base case**), and that every term after the 28th is T also (using step 2, the recursive part or **inductive case**). Here is what the proof would actually look like.

Proof. Let $P(n)$ be the statement "it is possible to make exactly n cents of postage using 5-cent and 8-cent stamps." We will show $P(n)$ is true for all $n \geq 28$.

First, we show that $P(28)$ is true: $28 = 4. 5 + 1. 8$, so we can make 28 cents using four 5-cent stamps and one 8-cent stamp.

Now suppose $P(k)$ is true for some arbitrary $k \geq 28$. Then it is possible to make k cents using 5-cent and 8-cent stamps. Note that since $k \geq 28$, it cannot be that we use fewer than three 5-cent stamps and fewer than three 8-cent stamps: using two of each would give only 26 cents. Now if we have made k cents using at least three 5-cent stamps, replace three 5-cent stamps by two 8-cent stamps. This replaces 15 cents of postage with 16 cents, moving from a total of k cents to k + 1 cents. Thus $P(k + 1)$ is true. On the other hand, if we have made k cents using at least three 8-cent stamps, then we can replace three 8-cent stamps with five 5-cent stamps, moving from 24 cents to 25 cents, giving a total of k + 1 cents of postage. So in this case as well $P(k + 1)$ is true.

Therefore, by the principle of mathematical induction, $P(n)$ is true for all $n \geq 28$.

Formalizing Proofs

What we did in the stamp example above works for many types of problems. Proof by induction is useful when trying to prove statements about all natural numbers, or all natural numbers greater than some fixed first case (like 28 in the example above), and in some other situations too. In particular, induction should be used when there is some way to go from one case to the next – when you can see how to always "do one more."

This is a big idea. Thinking about a problem *inductively* can give new insight into the problem. For example, to really understand the stamp problem, you should think about how any amount of postage (greater than 28 cents) can be made (this is non-inductive reasoning) and also how the ways in which postage can be made changes as the amount increases (inductive reasoning). When you are asked to provide a proof by induction, you are being asked to think about the problem dynamically; how does increasing *n* change the problem?

But there is another side to proofs by induction as well. In mathematics, it is not enough to understand a problem, you must also be able to communicate the problem to others. Like any discipline, mathematics has standard language and style, allowing mathematicians to share their ideas efficiently. Proofs by induction have a certain formal style, and being able to write in this style is important. It allows us to keep our ideas organized and might even help us with formulating a proof. Here is the general structure of a proof by mathematical induction:

Induction Proof Structure

Start by saying what the statement is that you want to prove: "Let $P(n)$ be the statement. . . " To prove that $P(n)$ is true for all $n \geq 0$, you must prove two facts:

- **Base case:** Prove that $P(0)$ is true. You do this directly. This is often easy.
- **Inductive case:** Prove that $P(k) \rightarrow P(k + 1)$ for all $k \geq 0$. That is, prove that for any $k \geq 0$ if $P(k)$ is true, then $P(k + 1)$ is true as well. This is the proof of an if . . . then . . . statement, so you can assume $P(k)$ is true ($P(k)$ is called the *inductive hypothesis*).

You must then explain why $P(k + 1)$ is also true, given that assumption.

Assuming you are successful on both parts above, you can conclude, "Therefore by the principle of mathematical induction, the statement P(n) is true for all $n \geq 0$."

Sometimes the statement P (n) will only be true for values of $n \geq 4$, for example, or some other value. In such cases, replace all the 0's above with 4's (or the other value).

The other advantage of formalizing inductive proofs is it allows us to verify that the logic behind this style of argument is valid. Why does induction work? Think of a row of dominoes set up standing on their edges. We want to argue that in a minute, all the dominoes will have fallen down. For this to happen, you will need to push the first domino. That is the base case. It will also have to be that the dominoes are close enough together that when any particular domino falls, it will cause the next domino to fall. That is the inductive case. If both of these conditions are met, you push the first domino over and each domino will cause the next to fall, then all the dominoes will fall.

Induction is powerful! Think how much easier it is to knock over dominoes when you don't have to push over each domino yourself. You just start the chain reaction, and the rely on the relative nearness of the dominoes to take care of the rest.

Think about our study of sequences. It is easier to find recursive definitions for sequences than closed formulas. Going from one case to the next is easier than going directly to a particular case. That is what is so great about induction. Instead of going directly to the (arbitrary) case for n, we just need to say how to get from one case to the next.

When you are asked to prove a statement by mathematical induction, you should first think about *why* the statement is true, using inductive reasoning. Explain why induction is the right thing to do, and roughly why the inductive case will work. Then, sit down and write out a careful, formal proof using the structure above.

Examples

Here are some examples of proof by mathematical induction.

Example 3.5.1

Prove for each natural number $n \geq 1$ that $1 + 2 + 3 + \cdots + n = \frac{n(n+1)}{2}$.

Solution

First, let's think inductively about this equation. In fact, we know this is true for other reasons (reverse and add comes to mind). But why might induction be applicable? The left-hand side adds up the numbers from 1 to n. If we know how to do that, adding just one more term (n + 1) would not be that hard. For example, if n = 100, suppose we know that the sum of the first 100 numbers is 5050 (so 1 + 2 + 3 + ... + 100 = 5050, which is true). Now to find the sum of the first 101 numbers, it makes more sense to just add 101 to 5050, instead of computing the entire sum again. We would have 1 + 2 + 3 + ... + 100 + 101 = 5050 + 101 = 5151. In fact, it would always be easy to add just one more term. This is why we should use induction.

Now the formal proof:

Proof. Let $P(n)$ be the statement $1 + 2 + 3 + \cdots + n = \frac{n(n+1)}{2}$. We will show that $P(n)$ is true for all natural numbers $n \geq 1$.

Base case: $P(1)$ is the statement $1 = \frac{1(1+1)}{2}$ which is clearly true.

Inductive case: Let $k \geq 1$ be a natural number. Assume (for induction) that P k is true. That means $1 + 2 + 3 + \cdots + k = \frac{k(k+1)}{2}$. We will prove that $P(k + 1)$ is true as well. That is, we must prove $1 + 2 + 3 + \cdots + k + (k + 1) = \frac{(k+1)(k+2)}{2}$. To prove this equation, start by adding k + 1 to both sides of the inductive hypothesis:

$$1 + 2 + 3 + \cdots + k + (k + 1) = \frac{k(k + 1)}{2} + (k + 1).$$

Now, simplifying the right side we get:

$$\frac{k(k + 1)}{2} + k + 1 = \frac{k(k + 1)}{2} + \frac{2(k + 1)}{2}$$

$$= \frac{k(k + 1) + 2(k + 1)}{2}$$

$$= \frac{(k + 2)(k + 1)}{2}.$$

Thus $P(k + 1)$ is true, so by the principle of mathematical induction $P(n)$ is true for all natural numbers $n \geq 1$.

Note that in the part of the proof in which we proved $P(k + 1)$ from $P(k)$, we used the equation P (k). This was the inductive hypothesis. Seeing how to use the inductive hypotheses is usually straight forward when proving a fact about a sum like this. In other proofs, it can be less obvious where it fits in.

Example 3.5.2

Prove that for all $n \in \mathbb{N}$, $6^n - 1$ is a multiple of 5.

Solution

Again, start by understanding the dynamics of the problem. What does increasing n do? Let's try with a few examples. If n = 1, then yes, $6^1 - 1 = 5$ is a multiple of 5. What does incrementing n to 2 look like? We get $6^2 - 1 = 35$, which again is a multiple of 5. Next, $n = 3$: but instead of just finding $6^3 - 1$, what did the increase in n do? We will still subtract 1, but now we are multiplying by another 6 first. Viewed another way, we are multiplying a number which is one more than a multiple of 5 by 6 (because $6^2 - 1$ is a multiple of 5, so 6^2 is one more than a multiple of 5). What do numbers which are one more than a multiple of 5 look like? They must have last digit 1 or 6. What happens when you multiply such a number by 6? Depends on the number, but in any case, the last digit of the new number must be a 6. And then if you subtract 1, you get last digit 5, so a multiple of 5.

The point is, every time we multiply by just one more six, we still get a number with last digit 6, so subtracting 1 gives us a multiple of 5. Now the formal proof:

Proof. Let P (n) be the statement, "$6^n - 1$ is a multiple of 5." We will prove that P (n) is true for all $n \in \mathbb{N}$.

Base case: $P(0)$ is true: $6^0 - 1 = 0$ which is a multiple of 5

Inductive case: Let k be an arbitrary natural number. Assume, for induction, that $P(k)$ is true. That is, $6^k - 1$ is a multiple of 5. Then $6^k - 1 = 5j$ for some integer j. This means that $6^k - = 5j + 1$. Multiply both sides by 6:

$$6^{k+1} = 6(5j + 1) = 30j + 6.$$

But we want to know about 6k+1 - 1, so subtract 1 from both sides:

$$6^{k+1} - 1 = 30j + 5.$$

Of course $30j + 5 = 5(6j + 1)$, so is a multiple of 5.

Therefore $6^{k+1} - 1$ is a multiple of 5, or in other words, $P(k + 1)$ is true. Thus, by the principle of mathematical induction $P(n)$ is true for all $n \in \mathbb{N}$.

We had to be a little bit clever (i.e., use some algebra) to locate the $6^k - 1$ inside of $6^{k+1} - 1$ before we could apply the inductive hypothesis. This is what can make inductive proofs challenging.

In the two examples above, we started with n = 1 or n = 0. We can start later if we need to.

Example 3.5.3

Prove that $n^2 < 2^n$ for all integers $n \geq 5$.

Solution

First, the idea of the argument. What happens when we increase n by 1? On the left-hand side, we increase the base of the square and go to the next square number. On the right-hand side, we increase the power of 2. This means we double the number. So the question is, how does doubling a number relate to increasing to the next square? Think about what the difference of two consecutive squares looks like. We have $(n + 1)^2 - n^2$. This factors:

$$(n+1)^2 - n^2 = (n+1-n)(n+1+n) = 2n+1.$$

But doubling the right-hand side increases it by 2^n, since $2^{n+1} = 2^n + 2^n$. When n is large enough, $2^n > 2n + 1$.

What we are saying here is that each time n increases, the left hand side grows by less than the right-hand side. So if the left-hand side starts smaller (as it does when $n = 5$), it will never catch up. Now the formal proof:

Proof. Let P(n) be the statement $n^2 < 2^n$. We will prove *P*(*n*) is true for all integers $n \geq 5$.

Base case: P(5) is the statement $5^2 < 2^5$. Since $5^2=25$ and $2^5 = 32$, we see that P (5) is indeed true.

Inductive case: Let $k \geq 5$ be an arbitrary integer. Assume, for induction, that P (k) is true. That is, assume $k^2 < 2^k$. We will prove that *P*(*k* + 1) is true, i.e., $(k + 1)^2 < 2^{k+1}$. To prove such an inequality, start with the left-hand side and work towards the right-hand side:

$$(k+1)^2 = k^2 + 2k + 1$$

$$\begin{aligned} &< 2^k + 2k + 1 && \text{... by the inductive hypothesis.} \\ &< 2^k + 2^k && \text{... since } 2k+1 < 2^k \text{ for } k \geq 5. \\ &= 2^{k+1}. \end{aligned}$$

Following the equalities and inequalities through, we get $(k +1)^2 < 2^{k+1}$, in other words, P(k + 1). Therefore by the principle of mathematical induction, *P*(*n*) is true for all $n \geq 5$.

The previous example might remind you of the racetrack principle from calculus, which says that if $f(a) < g(a)$, and $f'(x) < g'(x)$ for $x > a$, then $f(x) < g(x)$ for $x > a$. Same idea: the larger function is increasing at a faster rate than the smaller function, so the larger function will stay larger. In discrete math, we don't have derivatives, so we look at differences. Thus induction is the way to go.

Warning

With great power, comes great responsibility. Induction isn't magic. It seems very powerful to be able to assume P (k) is true. After all, we are trying to prove *P*(*n*) is true and the only difference is in the variable: k vs. n. Are we assuming that what we want to prove is true? Not really. We assume P (k) is true only for the sake of proving that P (k + 1) is true.

Still you might start to believe that you can prove anything with induction. Consider this incorrect "proof" that every Canadian has the same eye color: Let P (n) be the statement that any n Canadians have the same eye color. P (1) is true, since everyone has the same eye color as themselves. Now assume $P(k)$ is true. That is, assume that in any group of k Canadians, everyone has the same eye color. Now consider an arbitrary group of k + 1 Canadians. The first k of these must all have the same eye color, since P (k) is true. Also, the last k of these must have the same eye color, since $P(k)$ is true. So in fact, everyone the group must have the same eye color. Thus $P(k + 1)$ is true. So by the principle of mathematical induction, P (n) is true for all n.

Clearly something went wrong. The problem is that the proof that P(k) implies $P(k + 1)$ assumes that k ≥ 2. We have only shown $P(1)$ is true. In fact, $P(2)$ is false.

Strong Induction

Investigate!

Start with a square piece of paper. You want to cut this square into smaller squares, leaving no waste (every piece of paper you end up with must be a square). Obviously it is possible to cut the square into 4 squares. You can also cut it into 9 squares. It turns out you can cut the square into 7 squares (although not all the same size). What other numbers of squares could you end up with?

Attempt the above activity before proceeding

Sometimes, to prove that $P(k + 1)$ is true, it would be helpful to know that $P(k)$ and $P(k - 1)$ and $P(k - 2)$ are all true. Consider the following puzzle:

You have a rectangular chocolate bar, made up of n identical squares of chocolate. You can take such a bar and break it along any row or column. How many times will you have to break the bar to reduce it to n single chocolate squares?

At first, this question might seem impossible. Perhaps I meant to ask for the *smallest* number of breaks needed? Let's investigate.

Start with some small cases. If n = 2, you must have a 1 × 2 rectangle, which can be reduced to single pieces in one break. With $n = 3$, we must have a 1 × 3 bar, which requires two breaks: the first break creates a single square and *a* 1 × 2 bar, which we know takes one (more) break.

What about $n = 4$? Now we could have a 2 × 2 bar, or a 1 × 4 bar. In the first case, break the bar into two 2 2 bars, each which require one more break (that's a total of three breaks required). If we started with a 1 × 4 bar, we have choices for our first break. We could break the bar in half, creating two 1× 2 bars, or we could break off a single square, leaving a 1× 3 bar. But either way, we still need two more breaks, giving a total of three.

It is starting to look like no matter how we break the bar (and no matter how the n squares are arranged into a rectangle), we will always have the same number of breaks required. It also looks like that number is one less than *n*:

Conjecture 3.5.4

Given a n-square rectangular chocolate bar, it always takes n - 1 breaks to reduce the bar to single squares.

It makes sense to prove this by induction because after breaking the bar once, you are left with smaller chocolate bars. Reducing to *smaller* cases is what induction is all about. We can inductively assume we already know how to deal with these smaller bars. The problem is, if we are trying to prove the inductive case about a (k + 1) -square bar, we don't know that after the first break the remaining bar will have k squares. So we really need to assume that our conjecture is true for all cases less than k + 1.

Is it valid to make this stronger assumption? Remember, in induction we are attempting to prove that *P*(*n*) is true for all n. What if that were not the case? Then there would be some first n_0 for which $P(n_0)$ was false. Since n_0 is the first counterexample, we know that *P*(*n*) is true for all $n < n_0$. Now we proceed to prove that P n0 is actually true, based on the assumption that *P*(n) is true for all smaller n.

This is quite an advantage: we now have a stronger inductive hypothesis. We can assume that *P*(1) , *P*(2) , *P*(3) , . . . *P*(k) is true, just to show that *P*(k + 1) is true. Previously, we just assumed P k for this purpose.

It is slightly easier if we change our variables for strong induction.

Here is what the formal proof would look like:

Strong Induction Proof Structure.

Again, start by saying what you want to prove: "Let *P*(*n*) be the statement. . . " Then establish two facts:

- **Base case:** Prove that *P*(0) is true.
- **Inductive case:** Assume *P*(k) is true for all k < n. Prove that P(n) is true.

Conclude, "therefore, by strong induction, P(n) is true for all n > 0."

Of course, it is acceptable to replace 0 with a larger base case if needed.5 Let's prove our conjecture about the chocolate bar puzzle:

Proof. Let *P*(*n*) be the statement, "it takes n– 1 breaks to reduce a n-square chocolate bar to single squares."

- **Base case:** Consider $P(2)$. The squares must be arranged into a 1×2 rectangle, and we require $2-1=1$ breaks to reduce this to single squares.
- **Inductive case:** Fix an arbitrary $n \geq 2$ and assume $P(k)$ is true for all $k < n$. Consider a n-square rectangular chocolate bar. Break the bar once along any row or column. This results in two chocolate bars, say of sizes a and b. That is, we have an a-square rectangular chocolate bar, a b-square rectangular chocolate bar, and $a + b = n$.

We also know that $a < n$ and $b < n$, so by our inductive hypothesis, P(a) and P(b) are true. To reduce the a-square bar to single squares takes $a-1$ breaks; to reduce the b-square bar to single squares takes $b-1$ breaks. Doing this results in our original bar being reduced to single squares. All together it took the initial break, plus the a^{-1} and b^{-1} breaks, for a total of $1 + a - 1 + b - 1 = a + b - 1 = n - 1$ breaks. Thus P(n) is true.

Therefore, by strong induction, $P(n)$ is true for all $n \geq 2$.

Here is a more mathematically relevant example:

Example 3.5.5

Prove that any natural number greater than 1 is either prime or can be written as the product of primes.

Solution

First, the idea: if we take some number n, maybe it is prime. If so, we are done. If not, then it is composite, so it is the product of two smaller numbers. Each of these factors is smaller than n (but at least 2), so we can repeat the argument with these numbers. We have reduced to a smaller case.

Now the formal proof:

Proof. Let $P(n)$ be the statement, "n is either prime or can be written as the product of primes." We will prove $P(n)$ is true for all $n \geq 2$.

- **Base case:** $P(2)$ is true because 2 is indeed prime.
- **Inductive case:** assume $P(k)$ is true for all $k < n$. We want to show that P(n) is true. That is, we want to show that n is either prime or is the product of primes. If n is prime, we are done. If not, then n has more than 2 divisors, so we can write $n = m_1 . m_2$, with m_1 and m_2 less than n (and greater than 1). By the inductive hypothesis, m_1 and m_2 are each either prime or can be written as the product of primes. In either case, we have that n is written as the product of primes.

Thus by the strong induction, $P(n)$ is true for all $n \geq 2$.

Whether you use regular induction or strong induction depends on the statement you want to prove. If you wanted to be safe, you could always use strong induction. It really is stronger, so can accomplish everything "weak" induction can. That said, using regular induction is often easier since there is only one place you can use the induction hypothesis. There is also something to be said for elegance in proofs. If you can prove a statement using simpler tools, it is nice to do so.

As a final contrast between the two forms of induction, consider once more the stamp problem. Regular induction worked by showing how to increase postage by one cent (either replacing three 5-cent stamps with two 8-cent stamps, or thre stamps with five 5-cent stamps). We could give a slightly different proof using strong induction. First, we could show five base cases: it is possible to make 28, 29, 30, 31, and 32 cents (we would actually say how each of these is made). Now assume that it is possible to make k cents of postage for all $k < n$ as long as k 28. As long as $n > 32$, this means in particular we can make $k = n - 5$ cents. Now add a 5-cent stamp to get make *n* cents.

Exercises

1. On the way to the market, you exchange your cow for some magic dark chocolate espresso beans. These beans have the property that every night at midnight, each bean splits into two, effectively doubling your collection. You decide to take advantage of this and each morning (around 8am) you eat 5 beans.

   (a) Explain why it is true that if at noon on day n you have a number of beans ending in a 5, then at noon on day n + 1 you will still have a number of beans ending in a 5.

   (b) Why is the previous fact not enough to conclude that you will always have a number of beans ending in a 5? What additional fact would you need?

   (c) Assuming you have the additional fact in part (b), and have successfully proved the fact in part (a), how do you know that you will always have a number of beans ending in a 5? Illustrate what is going on by carefully explaining how the two facts above prove that you will have a number of beans ending in a 5 on *day* 4 specifically. In other words, explain why induction works in this context.

2. **Use induction to prove for all $n \in \mathbb{N}$ that** $\sum_{k=0}^{n} 2^k = 2^{n+1} - 1.$
3. **Prove that $7^n - 1$ is a multiple of 6 for all $n \in \mathbb{N}$.**
4. **Prove that $1 + 3 + 5 + \cdots + (2n - 1) - n^2$ for all $n \geq 1$.**
5. **Prove that $F_0 + F_2 + F_4 + \ldots\ldots + F_{2n} = F_{2n+1}$ 1 where F_n is the nth Fibonacci number.**
6. **Prove that $2^n < n!$ for all $n \geq 4$. (Recall, $n! = 1 \cdot 2 \cdot 3 \ldots\ldots\ldots n$.)**
7. **Prove, by mathematical induction, that $F_0 + F_1 + F_2 + \ldots\ldots\ldots + F_n = F_{n+2} - 1$, where F_n is the nth Fibonacci number ($F_0 = 0$, $F_1 = 1$ and $F_n = F_{n-1} + F_{n-2}$).**

8. Zombie Euler and Zombie Cauchy, two famous zombie mathematicians, have just signed up for Twitter accounts. After one day, Zombie Cauchy has more followers than Zombie Euler. Each day after that, the number of new followers of Zombie Cauchy is exactly the same as the number of new followers of Zombie Euler (and neither lose any followers). Explain how a proof by mathematical induction can show that on every day after the first day, Zombie Cauchy will have more followers than Zombie Euler. That is, explain what the base case and inductive case are, and why they together prove that Zombie Cauchy will have more followers on the 4th day.

9. Find the largest number of points which a football team cannot get exactly using just 3-point field goals and 7-point touchdowns (ignore the possibilities of safeties, missed extra points, and two point conversions). Prove your answer is correct by mathematical induction.

10. Prove that the sum of n squares can be found as follows

$$1^2 + 2^2 + 3^2 + \ldots + n^2 = \frac{n(n+1)(2n+1)}{6}.$$

11. Prove that the sum of the interior angles of a convex n-gon is $(n-2)\,180°$. (A convex n-gon is a polygon with n sides for which each interior angle is less than 180°.)

12. What is wrong with the following "proof" of the "fact" that $n + 3 = n + 7$ for all values of n (besides of course that the thing it is claiming to prove is false)?

    Proof. Let $P(n)$ be the statement that $n + 3 = n + 7$. We will prove that $P(n)$ is true for all $n \in \mathbb{N}$. Assume, for induction that $P(k)$ is true. That is, $k + 3 = k + 7$. We must show that $P(k + 1)$ is true. Now since $k + 3 = k + 7$, add 1 to both sides. This gives $k + 3 + 1 = k + 7 + 1$. Regrouping $(k + 1) + 3 = (k + 1) + 7$. But this is simply $P(k + 1)$. Thus by the principle of mathematical induction $P(n)$ is true for all $n \in \mathbb{N}$.

13. The proof in the previous problem does not work. But if we modify the "fact," we can get a working proof. Prove that $n + 3 < n + 7$ for all values of $n \in \mathbb{N}$. You can do this proof with algebra (without induction), but the goal of this exercise is to write out a valid induction proof.

14. Find the flaw in the following "proof" of the "fact" that $n < 100$ for every $n \in \mathbb{N}$.

    Proof. Let $P(n)$ be the statement $n < 100$. We will prove $P(n)$ is true for all $n \in \mathbb{N}$. First we establish the base case: when $n = 0$, $P(n)$ is true, because $0 < 100$. Now for the inductive step, assume $P(k)$ is true. That is, $k < 100$. Now if $k < 100$, then k is some number, like 80. Of course $80 + 1 = 81$ which is still less than 100. So $k + 1 < 100$ as well. But this is what $P(k + 1)$ claims, so we have shown that $P(k) \to P(k + 1)$. Thus by the principle of mathematical induction, $P(n)$ is true for all $n \in \mathbb{N}$.

15. While the above proof does not work (it better not since the statement it is trying to prove is false!) we can prove something similar. Prove that there is a strictly increasing sequence $a_1, a_2, a_3, \ldots$ of numbers (not necessarily integers) such that $a_n < 100$ for all $n \in \mathbb{N}$. (By strictly increasing we mean $a_n < a_{n+1}$ for all n. So each term must be larger than the last.)

16. What is wrong with the following "proof" of the "fact" that for all $n \in N$, the number $n^2 + n$ is odd?

    Proof. Let P(n) be the statement "$n^2 + n$ is odd." We will prove that P(n) is true for all $n \in N$. Suppose for induction that P(k) is true, that is, that $k^2 + k$ is odd. Now consider the statement $P(k + 1)$. Now $(k + 1)^2 + (k + 1) = k^2 + 2k + 1 + k + 1 = k^2 + k + 2k + 2$. By the inductive hypothesis, $k^2 + k$ is odd, and of course $2k + 2$ is even. An odd plus an even is always odd, so therefore $(k + 1)^2 + (k + 1)$ is odd. Therefore by the principle of mathematical induction, P(n) is true for all $n \in N$.

17. Now give a valid proof (by induction, even though you might be able to do so without using induction) of the statement, "for all $n \in N$, the number $n^2 + n$ is even."

18. Prove that there is a sequence of positive real numbers $a_0, a_1, a_2, \ldots$ such that the partial sum $a_0 + a_1 + a_2 + \ldots\ldots\ldots + a_n$ is strictly less than 2 for all $n \in N$. Hint: think about how you could define what a_{k+1} is to make the induction argument work.

19. Prove that every positive integer is either a power of 2, or can be written as the sum of distinct powers of 2.

20. Prove, using strong induction, that every natural number is either a Fibonacci number or can be written as the *sum* of *distinct* Fibonacci numbers.

21. Use induction to prove that if n people all shake hands with each other, that the total number of handshakes is $\frac{n(n-1)}{2}$.

22. Suppose that a particular real number x has the property that $x + \frac{1}{x}$ is an integer. Prove that $x^n + \frac{1}{x^n}$ is an integer for all natural numbers n.

23. Use induction to prove that $\sum_{k=0}^{n} \binom{n}{k} = 2^n$. That is, the sum of the nth row of Pascal's Triangle is 2^n.

24. Use induction to prove $\binom{4}{0} + \binom{5}{1} + \binom{6}{2} + \cdots + \binom{4+n}{n} = \binom{5+n}{n}$. (This is an example of the hockey stick theorem.)

25. Use the product rule for logarithms ($\log(ab) = \log(a) + \log(b)$) to provc, by induction on n, that $\log(a^n) = n \log(a)$, for all natural numbers $n \geq 2$.

26. Let $f1, f2, \ldots, f_n$ be differentiable functions. Prove, using induction, that

$$(f_1 + f_2 + \cdots + f_n)' = f_1' + f_2' + \cdots + f_n'.$$

You may assume $(f \cdot g)' = f' \cdot g'$ for any differentiable functions f and g.

27. **Suppose $f1, f2, \ldots, f_n$ are differentiable functions. Use mathematical induction to prove the generalized product rule:**

$$(f_1 f_2 f_3 \cdots f_n)' = f_1' f_2 f_3 \cdots f_n + f_1 f_2' f_3 \cdots f_n + f_1 f_2 f_3' \cdots f_n + \cdots + f_1 f_2 f_3 \cdots f_n'.$$

You may assume the product rule for two functions is true.

28. **You will prove that the Fibonacci numbers satisfy the identity $F^2_n + F^2_{n+1} = F_{2n+1}$. One way to do this is to prove the more general identity,**

$$F_m F_n + F_{m+1} F_{n+1} = F_{m+n+1},$$

and realize that when m = n we get our desired result.

Note that we now have two variables, so we want to prove this for all m ≥ 0 and all n ≥ 0 at the same time. For each such pair (*m*, *n*), let P(m, n) be the statement $F_m F_n + F_{m+1} F_{n+1} = F_{m+n+1}$

(a) First fix m = 0 and give a proof by mathematical induction that P(0, n) holds for all n ≥ 0. Note this proof will be very easy.

(b) Now fix an arbitrary n and give a proof by strong mathematical induction that P(m, n) holds for all m ≥ 0.

(c) You can now conclude that P (m, n) holds for all *m*, *n* ≥ 0. Do you believe that? Explain why this sort of induction is valid. For example, why do your proofs above guarantee that P (2, 3) is true?

29. **Given a square, you can cut the square into smaller squares by cutting along lines parallel to the sides of the original square (these lines do not need to travel the entire side length of the original square). For example, by cutting along the lines below, you will divide a square into 6 smaller squares:**

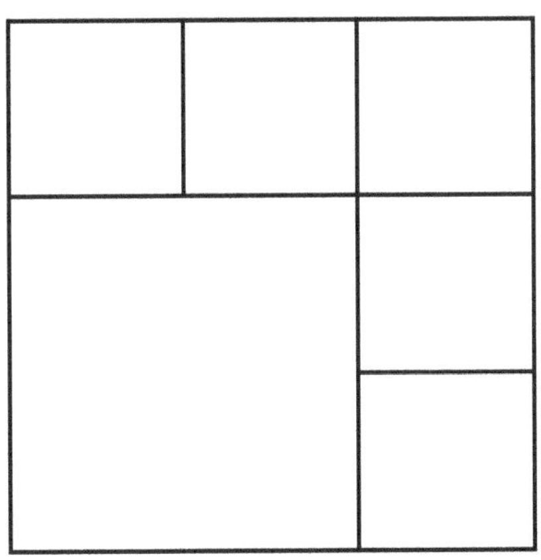

Prove, using strong induction, that it is possible to cut a square into *n* smaller squares for any *n* ≥ 6.

3.6 CONCISE SUMMARY

Investigate!

Each day your supply of magic chocolate covered espresso beans doubles (each one splits in half), but then you eat 5 of them. You have 10 at the start of day 0.

- Write out the first few terms of the sequence. Then give a recursive definition for the sequence and explain how you know it is correct.
- Prove, using induction, that the last digit of the number of beans you have on the nth day is always a 5 for all $n \geq 1$.
- Find a closed formula for the nth term of the sequence and prove it is correct by induction.

Attempt the above activity before proceeding

In this chapter we explored sequences and mathematical induction. At first these might not seem entirely related, but there is a link: recursive reasoning. When we have many cases (maybe infinitely many), it is often easier to describe a particular case by saying how it relates to other cases, instead of describing it absolutely. For sequences, we can describe the nth term in the sequence by saying how it is related to the *previous* term. When showing a statement involving the variable n is true for all values of n, we can describe why the case for n = k is true on the basis of why the case for $n = k^{-1}$ is true.

While thinking of problems recursively is often easier than thinking of them absolutely (at least after you get used to thinking in this way), our ultimate goal is to move beyond this recursive description. For sequences, we want to find *closed formulas* for the nth term of the sequence. For proofs, we want to know the statement is actually true for a particular n (not only under the assumption that the statement is true for the previous value of n). In this chapter we saw some methods for moving from recursive descriptions to absolute descriptions.

If the terms of a sequence increase by a constant difference or constant ratio (these are both recursive descriptions), then the sequences are arithmetic or geometric, respectively, and we have closed formulas for each of these based on the initial terms and common difference or ratio.

If the terms of a sequence increase at a polynomial rate (that is, if the differences between terms form a sequence with a polynomial closed formula), then the sequence is itself given by a polynomial closed formula (of degree one more than the sequence of differences).

If the terms of a sequence increase at an exponential rate, then we expect the closed formula for the sequence to be exponential. These sequences often have relatively nice recursive formulas, and the *characteristic root technique* allows us to find the closed formula for these sequences.

If we want to prove that a statement is true for all values of n (greater than some first small value), and we can describe why the statement being true for n = k implies the statement is true for n = k + 1, then the *principle of mathematical induction* gives us that the statement is true for all values of n (greater than the base case).

Throughout the chapter we tried to understand why these facts listed above are true. In part, that is what proofs, by induction or not, attempt to accomplish: they explain why mathematical truths are in fact truths. As we develop our ability to reason about mathematics, it is a good idea to make sure that the methods of our reasoning are sound. The branch of mathematics that deals with deciding whether reasoning is good or not is *mathematical logic,* the subject of the next chapter.

Chapter Review

1. **Find $3 + 7 + 11 + \cdots + 427$.**
2. **Consider the sequence $2, 6, 10, 14, \ldots, 4n + 6$.**
   (a) How many terms are there in the sequence?
   (b) What is the second-to-last term?
   (c) Find the sum of all the terms in the sequence.
3. **Consider the sequence given by $a_n = 2 \cdot 5^{n-1}$.**
   (a) Find the first 4 terms of the sequence.
   (b) What sort of sequence is this?
   (c) Find the *sum* of the first 25 terms. That is, compute $\sum_{k=1}^{25} a_k$.
4. **Consider the sequence 5, 11, 19, 29, 41, 55, ... Assume $a_1 = 5$.**
   (a) Find a closed formula for a_n, the *n*th term of the sequence, by writing each term as a sum of a sequence. Hint: first find a_0, but ignore it when collapsing the sum.
   (b) Find a closed formula again, this time using either polynomial fitting or the characteristic root technique (whichever is appropriate). Show your work.
   (c) Find a closed formula once again, this time by recognizing the sequence as a modification to some well known sequence(s). Explain.
5. **Use polynomial fitting to find a closed formula for the sequence $(a_n)_{n\geq 1}$:**

$$4, 11, 20, 31, 44, \ldots.$$

6. **Suppose the closed formula for a particular sequence is a degree 3 polynomial. What can you say about the closed formula for:**

(a) The sequence of partial sums.

(b) The sequence of second differences.

7. **Consider the sequence given recursively by $a_1 = 4$, $a_2 = 6$ and $a_n = a_{n-1} + a_{n-2}$.**

(a) Write out the first 6 terms of the sequence.

(b) Could the closed formula for an be a polynomial? Explain.

8. **The sequence $(a_n)_{n\geq1}$ starts -1, 0, 2, 5, 9, 14 . . . and has closed formula $a_n = \frac{(n+1)(n-2)}{2}$. Use this fact to find a closed formula for the sequence $(b_n)_{n\geq1}$ which starts 4, 10, 18, 28, 40, ...**

9. **The in song The Twelve Days of Christmas, my true love gave to me first 1 gift, then 2 gifts and 1 gift, then 3 gifts, 2 gifts and 1 gift, and so on. How many gifts did my true love give me all together during the twelve days?**

10. **Consider the recurrence relation an = 3an-1 + 10an-2 with first two terms $a_0 = 1$ and $a_1 = 2$.**

(a) Write out the first 5 terms of the sequence defined by this recurrence relation.

(b) Solve the recurrence relation. That is, find a closed formula for a_n.

11. **Consider the recurrence relation $a_n = 2a_{n-1} + 8a_{n-2}$, with initial terms $a_0 = 1$ and $a_1 = 3$.**

(a) Find the next two terms of the sequence (a_2 and a_3).

(b) Solve the recurrence relation. That is, find a closed formula for the nth term of the sequence.

12. **Your magic chocolate bunnies reproduce like rabbits: every large bunny produces 2 new mini bunnies each day, and each day every mini bunny born the previous day grows into a large bunny. Assume you start with 2 mini bunnies and no bunny ever dies (or gets eaten).**

(a) Write out the first few terms of the sequence.

(b) Give a recursive definition of the sequence and explain why it is correct.

(c) Find a closed formula for the nth term of the sequence.

13. **Consider the sequence of partial sums of squares of Fibonacci numbers: $F_1^2, F_1^2 + F_2^2, F_1^2 + F_2^2 + F_3^2, \ldots$ The sequences starts 1, 2, 6, 15, 40, . . .**

(a) Guess a formula for the nth partial sum, in terms of Fibonacci numbers. Hint: write each term as a product.

(b) Prove your formula is correct by mathematical induction.

(c) Explain what this problem has to do with the following picture:

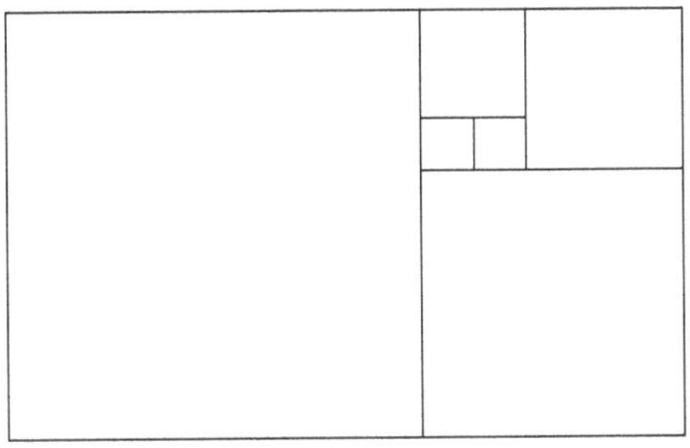

14. Prove the following statements by mathematical induction:

(a) $n! < n^n$ for $n \geq 2$

(b) $\frac{1}{1 \cdot 2} + \frac{1}{2 \cdot 3} + \frac{1}{3 \cdot 4} + \cdots + \frac{1}{n \cdot (n+1)} = \frac{n}{n+1}$ for all $n \in \mathbb{Z}^+$.

(c) $4^n - 1$ is a multiple of 3 for all $n \in \mathbb{N}$.

(d) The greatest amount of postage you cannot make exactly using 4 and 9 cent stamps is 23 cents.

(e) Every even number squared is divisible by 4.

15. Prove $1^3 + 2^3 + 3^3 + \cdots + n^3 = \left(\frac{n(n+1)}{2}\right)^2$ **holds for all $n \geq 1$, by mathematical induction.**

16. Suppose $a_0 = 1$, $a_1 = 1$ and $a_n = 3a_{n-1}$ $2a_{n-1}$. Prove, using strong induction, that $a_n = 1$ for all n.

17. Prove using induction that every set containing n elements has 2^n different subsets for any $n \geq 1$.

FOOTNOTES

1. This insight is usually attributed to Carl Friedrich Gauss, one of the greatest mathematicians of all time, who discovered it as a child when his unpleasant elementary teacher thought he would keep the class busy by requiring them to compute the lengthy sum.
2. Recurrence relations are sometimes called difference equations since they can describe the difference between terms and this highlights the relation to differential equations further.
3. You might or might not be familiar with these yet. We will consider these in Chapter 3.
4. This is not claiming that there are no amounts less than 27 cents which can also be made.
5. Technically, strong induction does not require you to prove a separate base

case. This is because when proving the inductive case, you must show that P(0) is true, assuming P(k) is true for all $k < 0$. But this is not any help so you end up proving P(0) anyway. To be on the safe side, we will always include the base case separately.

CHAPTER-4

SYMBOLIC LOGIC AND PROOFS

INTRODUCTION

Symbolic logic is the systematic study of the principles and methods used to distinguish valid reasoning from invalid reasoning. Unlike natural language, which can be ambiguous and imprecise, symbolic logic employs symbols and formal rules to represent logical relationships and arguments with clarity and rigor. At its core, symbolic logic enables us to analyze and manipulate propositions and arguments symbolically, facilitating precise deductions and proofs that serve as the foundation for mathematics, computer science, philosophy, and various scientific disciplines. This introduction explores the fundamental concepts of symbolic logic, its historical evolution, its application in constructing proofs, and its pivotal role in advancing our understanding of logical reasoning and problem-solving.

Logic is the study of consequence. Given a few mathematical statements or facts, we would like to be able to draw some conclusions. For example, if I told you that a particular real-valued function was continuous on the interval [0, 1] , and $f(0) = -1$ and $f(1) = 5$, can we conclude that there is some point between [0, 1] where the graph of the function crosses the x-axis? Yes, we can, thanks to the Intermediate Value Theorem from Calculus. Can we conclude that there is exactly one point? No. Whenever we find an "answer" in math, we really have a (perhaps hidden) argument. Mathematics is really about proving general statements (like the Intermediate Value Theorem), and this too is done via an argument, usually called a proof. We start with some given conditions, the premises of our argument, and from these we find a consequence of interest, our *conclusion.*

The problem is, as you no doubt know from arguing with friends, not all arguments are good arguments. A "bad" argument is one in which the conclusion does not follow from the premises, i.e., the conclusion is not a consequence of the premises. Logic is the study of what makes an argument good or bad. In other words, logic aims to determine in which cases a conclusion is, or is not, a consequence of a set of premises.

By the way, "argument" is actually a technical term in math (and philosophy, another discipline which studies logic):

Arguments

An **argument** is a set of statements, one of which is called the conclusion and the rest of which are called **premises.** An argument is said to be **valid** if the conclusion must be true whenever the premises are all true. An argument is **invalid** if it is not valid; it is possible for all the premises to be true and the conclusion to be false.

For example, consider the following two arguments:

If Edith eats her vegetables, then she can have a cookie.

Edith eats her vegetables.

\ Edith gets a cookie.

Florence must eat her vegetables in order to get a cookie.

Florence eats her vegetables.

\ Florence gets a cookie.

(The symbol means "therefore")

Are these arguments valid? Hopefully you agree that the first one is but the second one is not. Logic tells us why by analyzing the structure of the statements in the argument. Notice the two arguments above look almost identical. Edith and Florence both eat their vegetables. In both cases there is a connection between the eating of vegetables and cookies. But we claim that it is valid to conclude that Edith gets a cookie, but not that Florence does. The difference must be in the connection between eating vegetables and getting cookies. We need to be skilled at reading and comprehending these sentences. Do the two sentences mean the same thing? Unfortunately, in everyday language we are often sloppy, and you might be tempted to say they are equivalent. But notice that just because Florence must eat her vegetables, we have not said that doing so would be enough (she might also need to clean her room, for example). In everyday (non-mathematical) practice, you might be tempted to say this "other direction" is implied. In mathematics, we never get that luxury.

Before proceeding, it might be a good idea to quickly review Section 0.2 where we first encountered statements and the various forms they can take. The goal now is to see what mathematical tools we can develop to better analyze these, and then to see how this helps read and write proofs.

4.1 PROPOSITIONAL LOGIC

Investigate!

You stumble upon two trolls playing Stratego®. They tell you:

Troll 1: If we are cousins, then we are both knaves.

Troll 2: We are cousins or we are both knaves.

Could both trolls be knights? Recall that all trolls are either always-truth-telling knights or always-lying knaves.

Attempt the above activity before proceeding

A **proposition** is simply a statement. **Propositional logic** studies the ways statements can interact with each other. It is important to remember that propositional logic does not really care about the content of the statements. For example, in terms of propositional logic, the claims, "if the moon is made of cheese then basketballs are round," and "if spiders have eight legs then Sam walks with a limp" are exactly the same. They are both implications: statements of the form, $P \rightarrow Q$

Truth Tables

Here's a question about playing Monopoly:

If you get more doubles than any other player then you will lose, or if you lose then you must have bought the most properties.

True or false? We will answer this question, and won't need to know anything about Monopoly. Instead we will look at the logical form of the statement.

We need to decide when the statement $(P \rightarrow Q) \vee (Q \rightarrow R)$ is true. Using the definitions of the connectives in Section 0.2, we see that for this to be true, either $P \rightarrow Q$ must be true or $Q \rightarrow R$ must be true (or both). Those are true if either P is false or Q is true (in the first case) and Q is false or R is true (in the second case). So—yeah, it gets kind of messy. Luckily, we can make a chart to keep track of all the possibilities. Enter **truth tables.** The idea is this: on each row, we list a possible combination of T's and F's (for true and false) for each of the sentential variables, and then mark down whether the statement in question is true or false in that case. We do this for every possible combination of T's and F's. Then we can clearly see in which cases the statement is true or false. For complicated statements, we will first fill in values for each part of the statement, as a way of breaking up our task into smaller, more manageable pieces.

Since the truth value of a statement is completely determined by the truth values of its parts and how they are connected, all you really need to know is the truth tables for each of the logical connectives. Here they are:

P	Q	$P \wedge Q$
T	T	T
T	F	F
F	T	F
F	F	F

P	Q	$P \vee Q$
T	T	T
T	F	T
F	T	T
F	F	F

P	Q	$P \rightarrow Q$
T	T	T
T	F	F
F	T	T
F	F	T

P	Q	$P \leftrightarrow Q$
T	T	T
T	F	F
F	T	F
F	F	T

The truth table for negation looks like this:

P	$\neg P$
T	F
F	T

None of these truth tables should come as a surprise; they are all just restating the definitions of the connectives. Let's try another one.

Example 4.1.1

Make a truth table for the statement $\neg P \vee Q$.

Solution

Note that this statement is not $\neg(P \vee Q)$, the negation belongs to P alone. Here is the truth table:

P	Q	$\neg P$	$\neg P \vee Q$
T	T	F	T
T	F	F	F
F	T	T	T
F	F	T	T

We added a column for $\neg P$ to make filling out the last column easier. The entries in the ¬P column were determined by the entries in the P column. Then to fill in the final column, look only at the column for Q and the column for $\neg P$ and use the rule for $\vee$.

Now let's answer our question about monopoly:

Example 4.1.2

Analyze the statement, "if you get more doubles than any other player you will lose, or that if you lose you must have bought the most properties," using truth tables.

Solution

Represent the statement in symbols as $(P \to Q) \vee (Q \to R)$, where P is the statement "you get more doubles than any other player," Q is the statement "you will lose," and R is the statement "you must have bought the most properties." Now make a truth table.

The truth table needs to contain 8 rows in order to account for every possible combination of truth and falsity among the three statements. Here is the full truth table:

P	Q	R	$P \to Q$	$Q \to R$	$(P \to Q) \vee (Q \to R)$
T	T	T	T	T	T
T	T	F	T	F	T
T	F	T	F	T	T
T	F	F	F	T	T
F	T	T	T	T	T
F	T	F	T	F	T
F	F	T	T	T	T
F	F	F	T	T	T

The first three columns are simply a systematic listing of all possible combinations of T and F for the three statements (do you see how you would list the 16 possible combinations for four statements?). The next two columns are determined by the values of P, Q, and R and the definition of implication. Then, the last column is determined by the values in the previous two columns and the definition of $\vee$. It is this final column we care about.

Notice that in each of the eight possible cases, the statement in question is true. So our statement about monopoly is true (regardless of how many properties you own, how many doubles you roll, or whether you win or lose).

The statement about monopoly is an example of a **tautology**, a statement which is true on the basis of its logical form alone. Tautologies are always true but they don't tell us much about the world. No knowledge about monopoly was required to determine that the statement was true. In fact, it is equally true that "If the moon is made of cheese, then Elvis is still alive, or if Elvis is still alive, then unicorns have 5 legs."

Logical Equivalence

You might have noticed in **Example 4.1.1** that the final column in the truth table for $\neg P \vee Q$ is identical to the final column in the truth table for $P \to Q$:

P	Q	$P \to Q$	$\neg P \vee Q$
T	T	T	T
T	F	F	F
F	T	T	T
F	F	T	T

This says that no matter what P and Q are, the statements ¬P ∨ Q and P → Q either both true or both false. We therefore say these statements are logically equivalent.

Logical Equivalence

Two (molecular) statements P and Q are **logically equivalent** provided P is true precisely when Q is true. That is, P and Q have the same truth value under any assignment of truth values to their atomic parts.

To verify that two statements are logically equivalent, you can make a truth table for each and check whether the columns for the two statements are identical.

Recognizing two statements as logically equivalent can be very helpful. Rephrasing a mathematical statement can often lend insight into what it is saying, or how to prove or refute it. By using truth tables we can systematically verify that two statements are indeed logically equivalent.

Example 4.1.3

Are the statements, "it will not rain or snow" and "it will not rain and it will not snow" logically equivalent?

Solution

We want to know whether ¬(P ∨Q) is logically equivalent to ¬P ∧ ¬Q. Make a truth table which includes both statements:

P	Q	$\neg(P \vee Q)$	$\neg P \wedge \neg Q$
T	T	F	F
T	F	F	F
F	T	F	F
F	F	T	T

Since in every row the truth values for the two statements are equal, the two statements are logically equivalent.

Notice that this example gives us a way to "distribute" a negation over a disjunction (an "or"). We have a similar rule for distributing over conjunctions ("and"s):

De Morgan's Laws

$\neg(P \wedge Q)$ is logically equivalent to $\neg P \vee \neg Q$.

$\neg(P \vee Q)$ is logically equivalent to $\neg P \wedge \neg Q$.

This suggests there might be a sort of "algebra" you could apply to statements (okay, there is: it is called Boolean algebra) to transform one statement into another. We can start collecting useful examples of logical equivalence, and apply them in succession to a statement, instead of writing out a complicated truth table.

Implications are Disjunctions

$P \rightarrow Q$ is logically equivalent to $\neg P \vee Q$.

Example: "If a number is a multiple of 4, then it is even" is equivalent to, "a number is not a multiple of 4 or (else) it is even."

With this and De Morgan's laws, you can take any statement and *simplify* it to the point where negations are only being applied to atomic propositions. Well, actually not, because you could get multiple negations stacked up. But this can be easily dealt with:

Double Negation

$\neg\neg P$ is logically equivalent to P.

Example: "It is not the case that c is not odd" means "c is odd."

Let's see how we can apply the equivalences we have encountered so far.

Example 4.1.4

Prove that the statements $\neg(P \rightarrow Q)$ and $P \wedge \neg Q$ are logically equivalent without using truth tables.

Solution

We want to start with one of the statements, and trans- form it into the other through a sequence of logically equivalent as a disjunction this is logically equivalent to

$$\neg(\neg P \vee Q).$$

Now apply DeMorgan's law to get

$$\neg\neg P \wedge \neg Q.$$

Finally, use double negation to arrive at $P \wedge \neg Q$

Notice that the above example illustrates that the negation of an implication is NOT an implication: it is a conjunction! We saw this before, in Section 0.2, but it is so important and useful, it warants a second blue box here:

Negation of an Implication

The negation of an implication is a conjunction:

$\neg(P \to Q)$ is logically equivalent to $P \wedge \neg Q$.

That is, the only way for an implication to be false is for the hypothesis to be true *AND* the conclusion to be false.

To verify that two statements are logically equivalent, you can use truth tables or a sequence of logically equivalent replacements. The truth table method, although cumbersome, has the advantage that it can verify that two statements are NOT logically equivalent.

Example 4.1.5

Are the statements $(P \vee Q) \to R$ and $(P \to R) \vee (Q \to R)$ logically equivalent?

Solution

Note that while we could start rewriting these statements with logically equivalent replacements in the hopes of transforming one into another, we will never be sure that our failure is due to their lack of logical equivalence rather than our lack of imagination. So instead, let's make a truth table:

P	Q	R	$(P \vee Q) \to R$	$(P \to R) \vee (Q \to R)$
T	T	T	T	T
T	T	F	F	F
T	F	T	T	T
T	F	F	F	T
F	T	T	T	T
F	T	F	F	T
F	F	T	T	T
F	F	F	T	T

Look at the fourth (or sixth) row. In this case, $(P \to R) \vee (Q \to R)$ is true, but $(P \vee Q) \to R$ is false. Therefore the statements are not logically equivalent.

While we don't have logical equivalence, it is the case that whenever $(P \vee Q) \to R$ is true, so is $(P \to R) \vee (Q \to R)$. This tells us that we can deduce $(P \to R) \vee (Q \to R)$ from $(P \vee Q)$ R, just not the reverse direction.

Deductions

Investigate!

Holmes owns two suits: one black and one tweed. He always wears either a tweed suit or sandals. Whenever he wears his tweed suit and a purple shirt, he chooses to not wear a tie. He never wears the tweed suit unless he is also wearing either a purple shirt or sandals. Whenever he wears sandals, he also wears a purple shirt. Yesterday, Holmes wore a bow tie. What else did he wear?

Attempt the above activity before proceeding

Earlier we claimed that the following was a valid argument:

If Edith eats her vegetables, then she can have a cookie. Edith ate her vegetables. Therefore Edith gets a cookie.

How do we know this is valid? Let's look at the form of the statements. Let P denote "Edith eats her vegetables" and Q denote "Edith can have a cookie." The logical form of the argument is then:

$$\begin{array}{rc} & P \to Q \\ & P \\ \hline \therefore & Q \end{array}$$

This is an example of a **deduction rule**, an argument form which is always valid. This one is a particularly famous rule called modus ponens. Are you convinced that it is a valid deduction rule? If not, consider the following truth table:

P	Q	$P \to Q$
T	T	T
T	F	F
F	T	T
F	F	T

This is just the truth table for P→ Q, but what matters here is that all the lines in the deduction rule have their own column in the truth table. Remember that an argument is valid provided the conclusion must be true given that the premises are true. The premises in this case are P Q and P. Which rows of the truth table correspond to both of these being true? P is true in the first two rows, and of those, only the first row has P → Q true as well. And lo-and-behold, in this one case, Q is also true. So if P → Q and P are both true, we see that Q must be true as well. Here are a few more examples.

Example 4.1.6

Show that

$$\begin{array}{rc} & P \to Q \\ & \neg P \to Q \\ \hline \therefore & Q \end{array}$$

is a valid deduction rule.

Solution

We make a truth table which contains all the lines of the argument form:

P	Q	$P \to Q$	$\neg P$	$\neg P \to Q$
T	T	T	F	T
T	F	F	F	T
F	T	T	T	T
F	F	T	T	F

(we include a column for $\neg P$ just as a step to help getting the column for $\neg P \to Q$).

Now look at all the rows for which both $P \to Q$ and $\neg P \to Q$ are true. This happens only in rows 1 and 3. Hey! In those rows Q is true as well, so the argument form is valid (it is a valid deduction rule).

Example 4.1.7

Decide whether

$$\begin{array}{rc} & P \to R \\ & Q \to R \\ & R \\ \hline \therefore & P \vee Q \end{array}$$

is a valid deduction rule.

Solution

Let's make a truth table containing all four statements.

P	Q	R	$P \to R$	$Q \to R$	$P \vee Q$
T	T	T	T	T	T
T	T	F	F	F	T
T	F	T	T	T	T
T	F	F	F	T	T
F	T	T	T	T	T
F	T	F	T	F	T
F	F	T	T	T	F
F	F	F	T	T	F

Look at the second to last row. Here all three premises of the argument are true, but the conclusion is false. Thus this is not a valid deduction rule.

While we have the truth table in front of us, look at rows 1, 3, and 5. These are the only rows in which all of the statements $P \to R$, $Q \to R$, and $P \vee Q$ are true. It also happens that R is true in these rows as well. Thus we have discovered a new deduction rule we know *is* valid:

$$\begin{array}{rc} & P \to R \\ & Q \to R \\ & P \vee Q \\ \hline \therefore & R \end{array}$$

Beyond Propositions

As we saw in **Section 0.2,** not every statement can be analyzed using logical connectives alone. For example, we might want to work with the statement:

- All primes greater than 2 are odd.

To write this statement symbolically, we must use quantifiers. We can translate as follows:

$$\forall x((P(x) \wedge x > 2) \to O(x)).$$

In this case, we are using $P(x)$ to denote "x is prime" and $O(x)$ to denote "x is odd." These are not propositions, since their truth value depends on the input x. Better to think of P and O as denoting properties of their input. The technical term for these is **predicates** and when we study them in logic, we need to use **predicate logic.**

It is important to stress that predicate logic extends propositional logic (much in the way quantum mechanics extends classical mechanics). You will notice that our statement above still used the (propositional) logical connectives. Everything that we learned about logical equivalence and deductions still applies. However, predicate logic allows us to analyze statements at a higher resolution, digging down into the individual propositions P, Q, etc.

A full treatment of predicate logic is beyond the scope of this text. One reason is that there is no systematic procedure for deciding whether two statements in predicate logic are logically equivalent (i.e., there is no analogue to truth tables here). Rather, we end with a two examples of logical equivalence and deduction, to pique your interest.

Example 4.1.8

Suppose we claim that there is no smallest number. We can translate this into symbols as

$$\neg\exists x\forall y(x \leq y)$$

(literally, "it is not true that there is a number x such that for all numbers y, x is less than or equal to *y*").

However, we know how negation interacts with quantifiers: we can pass a negation over a quantifier by switching the quantifier type (between universal and existential). So the statement above should be *logically equivalent* to

$$\forall x\exists y(y < x).$$

Notice that $y < x$ is the negation of $x \leq y$. This literally says, "for every number x there is a number y which is smaller than x." We see that this is another way to make our original claim.

Example 4.1.9

Can you switch the order of quantifiers? For example, consider the two statements:

$$\forall x\exists yP(x, y) \text{ and } \exists y\forall xP(x, y).$$

Are these logically equivalent?

Solution

These statements are NOT logically equivalent. To see this, we should provide an interpretation of the predicate P (x, y) which makes one of the statements true and the other false.

Let *P(x, y)* be the predicate $x < y$. It is true, in the natural numbers, that for all x there is some y greater than that x (since there are infinitely many numbers). However, there is not a natural number y which is greater than every number x. Thus it is possible for $\forall x\exists yP(x, y)$ to be true while $\exists y\forall xP(x, y)$ is false.

We cannot do the reverse of this though. If there is some y for which every x satisfies P(x, y) , then certainly for every x there is some y which satisfies P(x, y) . The first is saying we can find one y that works for every x. The second allows different y's to work for different x's, but there is nothing preventing us from using the same y that work for every x. In other words, while we don't have logical equivalence between the two statements, we do have a valid deduction rule:

$$\begin{array}{rl} & \exists y\forall xP(x, y) \\ \hline \therefore & \forall x\exists yP(x, y) \end{array}$$

Put yet another way, this says that the single statement

$$\exists y \forall x P(x, y) \rightarrow \forall x \exists y P(x, y)$$

is always true. This is sort of like a tautology, although we reserve that term for necessary truths in propositional logic. A statement in predicate logic that is necessarily true gets the more prestigious designation of a **law of logic** (or sometimes **logically valid**, but that is less fun).

Exercises

1. **Consider the statement about a party, "If it's your birthday or there will be cake, then there will be cake."**
   (a) Translate the above statement into symbols. Clearly state which statement is P and which is Q.
   (b) Make a truth table for the statement.
   (c) Assuming the statement is true, what (if anything) can you conclude if there will be cake?
   (d) Assuming the statement is true, what (if anything) can you conclude if there will not be cake?
   (e) Suppose you found out that the statement was a lie. What can you conclude?
2. **Make a truth table for the statement $(P \vee Q) \rightarrow (P \wedge Q)$.**
3. **Make a truth table for the statement $\neg P \wedge (Q \rightarrow P)$. What can you conclude about P and Q if you know the statement is true?**
4. **Make a truth table for the statement $\neg P \rightarrow (Q \wedge R)$.**
5. **Geoff Poshingten is out at a fancy pizza joint, and decides to order a calzone. When the waiter asks what he would like in it, he replies, "I want either pepperoni or sausage. Also, if I have sausage, then I must also include quail. Oh, and if I have pepperoni or quail then I must also have ricotta cheese."**
   (a) Translate Geoff's order into logical symbols.
   (b) The waiter knows that Geoff is either a liar or a truth-teller (so either everything he says is false, or everything is true). Which is it?
   (c) What, if anything, can the waiter conclude about the ingredients in Geoff's desired calzone?
6. **Determine whether the following two statements are logically equivalent: $\neg(P \rightarrow Q)$ and $P \wedge \neg Q$. Explain how you know you are correct.**
7. **Are the statements $P \rightarrow (Q \vee R)$ and $(P \rightarrow Q) \vee (P \rightarrow R)$ logically equivalent?**
8. **Simplify the following statements (so that negation only appears right before variables).**

(a) $\neg(P \to \neg Q)$.

(b) $(\neg P \vee \neg Q) \to \neg(\neg Q \wedge R)$.

(c) $\neg((P \to \neg Q) \vee \neg(R \wedge \neg R))$.

(d) It is false that if Sam is not a man then Chris is a woman, and that Chris is not a woman.

9. Use De Morgan's Laws, and any other logical equivalence facts you know to simplify the following statements. Show all your steps. Your final statements should have negations only appear directly next to the sentence variables or predicates (P, Q, F (x), etc.), and no double negations. It would be a good idea to use only conjunctions, disjunctions, and negations.

(a) $\neg((\neg P \wedge Q) \vee \neg(R \vee \neg S))$.

(b) $\neg((\neg P \to \neg Q) \wedge (\neg Q \to R))$ (careful with the implications).

(c) For both parts above, verify your answers are correct using truth tables. That is, use a truth table to check that the given statement and your proposed simplification are actually logically equivalent.

10. Consider the statement, "If a number is triangular or square, then it is not prime"

(a) Make a truth table for the statement $(T \vee S) \to \neg P$.

(b) If you believed the statement was false, what properties would a counterexample need to possess? Explain by referencing your truth table.

(c) If the statement were true, what could you conclude about the number 5657, which is definitely prime? Again, explain using the truth table.

11. Tommy Flanagan was telling you what he ate yesterday afternoon. He tells you, "I had either popcorn or raisins. Also, if I had cucumber sandwiches, then I had soda. But I didn't drink soda or tea." Of course you know that Tommy is the worlds worst liar, and everything he says is false. What did Tommy eat? Justify your answer by writing all of Tommy's statements using sentence variables (*P, Q, R, S, T*), taking their negations, and using these to deduce what Tommy actually ate.

12. Determine if the following deduction rule is valid:

$$\begin{array}{rc} & P \vee Q \\ & \neg P \\ \hline \therefore & Q \end{array}$$

13. Determine if the following is a valid deduction rule:

$$\begin{array}{rc} & P \to (Q \vee R) \\ & \neg(P \to Q) \\ \hline \therefore & R \end{array}$$

14. Determine if the following is a valid deduction rule:

$$\begin{array}{rl} & (P \wedge Q) \rightarrow R \\ & \neg P \vee \neg Q \\ \hline \therefore & \neg R \end{array}$$

15. Can you chain implications together? That is, if P→ Q and Q→ R, does that means the P→ R? Can you chain more implications together? Let's find out:

Prove that the following is a valid deduction rule:

$$\begin{array}{rl} & P \rightarrow Q \\ & Q \rightarrow R \\ \hline \therefore & P \rightarrow R \end{array}$$

Prove that the following is a valid deduction rule for any n ≥ 2:

$$\begin{array}{rl} & P_1 \rightarrow P_2 \\ & P_2 \rightarrow P_3 \\ & \vdots \\ & P_{n-1} \rightarrow P_n \\ \hline \therefore & P_1 \rightarrow P_n. \end{array}$$

I suggest you don't go through the trouble of writing out a 2n row truth table. Instead, you should use part (a) and mathematical induction.

16. We can also simplify statements in predicate logic using our rules for passing negations over quantifiers, and then applying propositional logical equivalence to the "inside" propositional part. Simplify the statements below (so negation appears only directly next to predicates).

(a) ¬∃x∀y(¬O(x) ∨ E(y)).

(b) ¬∀x¬∀y¬(x < y ∧ ∃z(x < z ∨ y < z)).

(c) There is a number n for which no other number is either less n than or equal to n.

(d) It is false that for every number n there are two other numbers which n is between.

17. Simplify the statements below to the point that negation symbols occur only directly next to predicates.

(a) ¬∀x∀y(x < y ∨ y < x).

(b) ¬(∃xP(x) → ∀yP(y)).

18. Simplifying negations will be especially useful in the next section when we try to prove a statement by considering what would happen if it were false. For each statement below, write the *negation* of the statement as simply as possible. Don't just say, "it is false that . . . ".

(a) Every number is either even or odd.

(b) There is a sequence that is both arithmetic and geometric.

(c) For all numbers n, if n is prime, then n + 3 is not prime.v

19. Suppose P and Q are (possibly molecular) propositional statements. Prove that P and Q are logically equivalent if any only if P↔Q is a tautology.

20. Suppose P_1, P_2, . . . , P_n and Q are (possibly molecular) propositional statements. Suppose further that

$$\begin{array}{l} P_1 \\ P_2 \\ \vdots \\ P_n \\ \hline \therefore\ Q \end{array}$$

is a valid deduction rule. Prove that the statement

$(P_1 \wedge P_2 \wedge \cdots \wedge P_n) \to Q$ is a tautology.

4.2 PROOFS

Investigate!

Decide which of the following are valid proofs of the following statement:

If ab is an even number, then *a* or *b* is even.

Suppose *a* and *b* are odd. That is, a = 2k + 1 and b = 2m + 1 for some integers *k* and *m*. Then

$$\begin{aligned} ab &= (2k+1)(2m+1) \\ &= 4km + 2k + 2m + 1 \\ &= 2(2km + k + m) + 1. \end{aligned}$$

Therefore *ab* is odd.

Assume that *a* or *b* is even - say it is a (the case where b is even is identical). That is, *a* = 2*k* for some integer *k*. Then

$$\begin{aligned} ab &= (2k)b \\ &= 2(kb). \end{aligned}$$

Thus ab is even.

Suppose that ab is even but a and b are both odd. Namely, $ab = 2n$, $a = 2k + 1$ and $b = 2j + 1$ for some integers n, k, and j. Then

$$\begin{aligned} 2n &= (2k+1)(2j+1) \\ 2n &= 4kj + 2k + 2j + 1 \\ n &= 2kj + k + j + 0.5. \end{aligned}$$

But since $2kj + k + j$ is an integer, this says that the integer n is equal to a non-integer, which is impossible.

Let ab be an even number, say $ab = 2n$, and a be an odd number, say $a = 2k + 1$.

$$\begin{aligned} ab &= (2k+1)b \\ 2n &= 2kb + b \\ 2n - 2kb &= b \\ 2(n - kb) &= b. \end{aligned}$$

Therefore b must be even.

Attempt the above activity before proceeding

Anyone who doesn't believe there is creativity in mathematics clearly has not tried to write proofs. Finding a way to convince the world that a particular statement is necessarily true is a mighty undertaking and can often be quite challenging. There is not a guaranteed path to success in the search for proofs. For example, in the summer of 1742, a German mathematician by the name of Christian Goldbach wondered whether every even integer greater than 2 could be written as the sum of two primes. Centuries later, we still don't have a proof of this apparent fact (computers have checked that "Goldbach's Conjecture" holds for all numbers less than 4×10^{18}, which leaves only infinitely many more numbers to check).

Writing proofs is a bit of an art. Like any art, to be truly great at it, you need some sort of inspiration, as well as some foundational technique. Just as musicians can learn proper fingering, and painters can learn the proper way to hold a brush, we can look at the proper way to construct arguments. A good place to start might be to study a classic.

Theorem 4.2.1 *There are infinitely many primes.*

Proof. Suppose this were not the case. That is, suppose there are only finitely many primes. Then there must be a last, largest prime, call it p. Consider the number

$$N = p! + 1 = (p \cdot (p-1)\ 3 \cdot 2 \cdot 1) + 1.$$

Now N is certainly larger than p. Also, N is not divisible by any number less than or equal to p, since every number less than or equal to p divides p!. Thus the prime factorization of N contains prime numbers (possibly just N itself) all greater than p. So p is not the largest prime, a contradiction. Therefore there are infinitely many primes.

This proof is an example of a *proof by contradiction,* one of the standard styles of mathematical proof. First and foremost, the proof is an argument. It contains sequence of statements, the last being the *conclusion* which follows from the previous statements. The argument is valid so the conclusion must be true if the premises are true. Let's go through the proof line by line.

- Suppose there are only finitely many primes. [*this is a premise. Note the use of "suppose."*]
- There must be a largest prime, call it p. [*follows from line 1, by the definition of "finitely many."*]
- Let $N = p! + 1$. [*basically just notation, although this is the inspired part of the proof; looking at $p! + 1$ is the key insight.*]
- N is larger than p. [*by the definition of $p!$*]
- N is not divisible by any number less than or equal to p. [*by definition, $p!$ is divisible by each number less than or equal to p, so $p! + 1$ is not.*]
- The prime factorization of N contains prime numbers greater than p. [*since N is divisible by each prime number in the prime factorization of N, and by line 5.*]
- Therefore p is not the largest prime. [*by line 6, N is divisible by a prime larger than p.*]
- This is a contradiction. [*from line 2 and line 7: the largest prime is p and there is a prime larger than p.*]
- Therefore there are infinitely many primes. [*from line 1 and line 8: our only premise lead to a contradiction, so the premise is false.*]

We should say a bit more about the last line. Up through line 8, we have a valid argument with the premise "there are only finitely many primes" and the conclusion "there is a prime larger than the largest prime." This is a valid argument as each line follows from previous lines. So if the premises are true, then the conclusion must be true. However, the conclusion is NOT true. The only way out: the premise must be false.

The sort of line-by-line analysis we did above is a great way to really understand what is going on. Whenever you come across a proof in a textbook, you really should make sure you understand what each line is saying and why it is true. Additionally, it is equally important to understand the overall structure of the proof. This is where using tools from logic is helpful. Luckily there are a relatively small number of standard proof styles that keep showing up again and again. Being familiar with these can help understand proof, as well as give ideas of how to write your own.

Direct Proof

The simplest (from a logic perspective) style of proof is a **direct proof.** Often all that is required to prove something is a systematic explanation of what everything means. Direct proofs are especially useful when proving implications. The general format to prove $P \to Q$ is this:

Assume P. Explain, explain, . . . , explain. Therefore Q.

Often we want to prove universal statements, perhaps of the form $\forall x(P(x) \to Q(x))$. Again, we will want to assume P(x) is true and deduce Q(x) . But what about the x? We want this to work for all x. We accomplish this by fixing x to be an arbitrary element (of the sort we are interested in).

Here are a few examples. First, we will set up the proof structure for a direct proof, then fill in the details.

Example 4.2.2

Prove: For all integers n, if n is even, then n^2 is even.

Solution

The format of the proof will be this: Let n be an arbitrary integer. Assume that n is even. Explain explain explain. Therefore n^2 is even.

To fill in the details, we will basically just explain what it means for n to be even, and then see what that means for n^2. Here is a complete proof.

Proof. Let n be an arbitrary integer. Suppose n is even. Then $n = 2k$ for some integer k. Now $n^2 = (2k)^2 = 4k^2 = 2(2k^2)$. Since $2k^2$ is an integer, n^2 is even.

Example 4.2.3

Prove: For all integers *a, b,* and *c,* if $a|b$ and $b|c$ then $a|c$. (Here $x|y$, read "x divides y" means that y is a multiple of x, i.e., that x will divide into y without remainder).

Solution

Even before we know what the divides symbol means, we can set up a direct proof for this statement. It will go something like this: Let a, b, and c be arbitrary integers. Assume that $a|b$ and $b|c$. Dot dot dot. Therefore $a|c$.

How do we connect the dots? We say what our hypothesis ($a|b$ and ($a|c$) really means and why this gives us what the conclusion ($a|c$) really means. Another way to say that $a|b$ is to say that $b = ka$ for some integer k (that is, that b is a multiple of a). What are we going for? That $c = la$, for some integer l (because we want c to be a multiple of a). Here is the complete proof.

Proof. Let a, b, and c be integers. Assume that a |b and b |c. In other words, b is a multiple of a and c is a multiple of b. So there are integers k and j such that b = ka and c = jb. Combining these (through substitution) we get that $c = jka$. But jk is an integer, so this says that c is a multiple of a. Therefore a|c.

Proof by Contrapositive

Recall that an implication P → Q is logically equivalent to its contrapositive ¬Q → ¬P. There are plenty of examples of statements which are hard to prove directly, but whose contrapositive can easily be proved directly. This is all that **proof by contrapositive** does. It gives a direct proof of the contrapositive of the implication. This is enough because the contrapositive is logically equivalent to the original implication.

The skeleton of the proof of P→ Q by contrapositive will always look roughly like this:

Assume ¬Q. Explain, explain, . . . explain. Therefore ¬*P*.

As before, if there are variables and quantifiers, we set them to be arbitrary elements of our domain. Here are two examples:

Example 4.2.4

Is the statement "for all integers n, if n^2 is even, then n is even" true?

Solution

This is the converse of the statement we proved above using a direct proof. From trying a few examples, this statement definitely appears to be true. So let's prove it.

A direct proof of this statement would require fixing an arbitrary n and assuming that n^2 is even. But it is not at all clear how this would allow us to conclude anything about n. Just because $n^2 = 2k$ does not in itself suggest how we could write n as a multiple of 2.

Try something else: write the contrapositive of the statement. We get, for all integers n, if n is odd then n^2 is odd. This looks much more promising. Our proof will look something like this:

Let n be an arbitrary integer. Suppose that n is not even. This means that . . . . In other words But this is the same as saying . . . . Therefore n^2 is not even.

Now we fill in the details:

Proof. We will prove the contrapositive. Let n be an arbitrary integer. Suppose that n is not even, and thus odd. Then $n = 2k + 1$ for some integer k. Now $n^2 = (2k + 1)^2 = 4k^2 + 4k + 1 = 2(2k^2 + 2k) + 1$.

Since $2k^2 + 2k$ is an integer, we see that n^2 is odd and therefore not even.

Example 4.2.5

Prove: for all integers a and b, if $a + b$ is odd, then a is odd or b is odd.

Solution

The problem with trying a direct proof is that it will be hard to separate a and b from knowing something about $a + b$. On the other hand, if we know something about a and b separately, then combining them might give us information about a + b. The contrapositive of the statement we are trying to prove is: for all integers a and b, if a and b are even, then a + b is even. Thus our proof will have the following format:

Let a and b be integers. Assume that a and b are both even. la la la. Therefore a + b is even.

Here is a complete proof:

- *Proof.* Let a and b be integers. Assume that a and b are even. Then a = 2k and b = 2l for some integers k and l. Now a + b = 2k + 2l = $2(k + l)$. Since $k + l$ is an integer, we see that a + b is even, completing the proof.
- Note that our assumption that a and b are even is really the negation of a or b is odd. We used De Morgan's law here.
- We have seen how to prove some statements in the form of implications: either directly or by contrapositive. Some statements are not written as implications to begin with.

Example 4.2.6

Consider the following statement: for every prime number p, either p = 2 or p is odd. We can rephrase this: for every prime number p, if $p \neq 2$, then p is odd. Now try to prove it.

Solution

Proof. Let p be an arbitrary prime number. Assume p is not odd. So p is divisible by 2. Since p is prime, it must have exactly two divisors, and it has 2 as a divisor, so p must be divisible by only 1 and 2. Therefore $p = 2$. This completes the proof (by contrapositive).

Proof by Contradiction

There might be statements which really cannot be rephrased as implications.

For example, "$\sqrt{2}$ is irrational." In this case, it is hard to know where to start. What can we assume? Well, say we want to prove the statement P. What if we could prove that ¬P →Q where Q was false? If this implication is true, and Q is false, what can we say about P? It must be false as well, which makes P true!

This is why **proof by contradiction** works. If we can prove that¬ P leads to a contradiction, then the only conclusion is that *P* is false, so *P* is true. That's what we wanted to prove. In other words, if it is impossible for *P* to be false, *P* must be true.

Here are three examples of proofs by contradiction:

Example 4.2.7

Prove that $\sqrt{2}$ is irrational.

Solution

Proof. Suppose not. Then $\sqrt{2}$ is equal to a fraction $\frac{a}{b}$ · Without loss of generality, assume a is in lowest terms (otherwise reduce the fraction). So,

$$2 = \frac{a^2}{b^2}$$

$$2b^2 = a^2.$$

Thus a^2 is even, and as such a is even. So a = 2k for some integer k, and $a^2 = 4k^2$. We then have,

$$2b^2 = 4k^2$$

$$b^2 = 2k^2.$$

Thus b^2 is even, and as such b is even. Since a is also even, we see that $\frac{a}{b}$ is not in lowest terms, a contradiction. Thus $\sqrt{2}$ is irrational.

Example 4.2.8

Prove: There are no integers x and y such that $x^2 = 4y + 2$.

Solution

Proof. We proceed by contradiction. So suppose there are integers x and y such that $x^2 = 4y + 2 = 2(2y + 1)$. So x^2 is even. We have seen that this implies that x is even. So x = 2k for some integer k. Then $x^2 = 4k^2$. This in turn gives $2k^2 = (2y + 1)$. But $2k^2$ is even, and 2y + 1 is odd, so these cannot be equal. Thus we have a contradiction, so there must not be any integers x and y such that $x^2 = 4y + 2$.

Example 4.2.9

The **Pigeonhole Principle:** If more than *n* pigeons fly into n pigeon holes, then at least one pigeon hole will contain at least two pigeons. Prove this!

Solution

Proof. Suppose, contrary to stipulation, that each of the pigeon holes contain at most one pigeon. Then at most, there will be n pigeons. But we assumed that there are more than n pigeons, so this is impossible. Thus there must be a pigeonhole with more than one pigeon.

While we phrased this proof as a proof by contradiction, we could have also used a proof by contrapositive since our contradiction was simply the negation of the hypothesis. Sometimes this will happen, in which case you can use either style of proof. There are examples however where the contradiction occurs "far away" from the original statement.

Proof by (Counter) Example

It is almost NEVER okay to prove a statement with just an example. Certainly none of the statements proved above can be proved through an example. This is because in each of those cases we are trying to prove that something holds of all integers. We claim that n^2 being even implies that n is even, no matter what integer n we pick. Showing that this works for n = 4 is not even close to enough.

This cannot be stressed enough. If you are trying to prove a statement of the form $\forall x P(x)$, you absolutely CANNOT prove this with an example.1 However, existential statements can be proven this way. If we want to prove that there is an integer n such that n^2 n + 41 is not prime, all we need to do is find one. This might seem like a silly thing to want to prove until you try a few values for *n*.

n	1	2	3	4	5	6	7
$n^2 - n + 41$	41	43	47	53	61	71	83

So far we have gotten only primes. You might be tempted to conjecture, "For all positive integers n, the number $n^2 - n + 41$ is prime." If you wanted to prove this, you would need to use a direct proof, a proof by contrapositive, or another style of proof, but certainly it is not enough to give even 7 examples. In fact, we can prove this conjecture is false by proving its negation: "There is a positive integer n such that $n^2 - n + 41$ is not prime." Since this is an existential statement, it suffices to show that there does indeed exist such a number.

In fact, we can quickly see that n = 41 will give 41^2 which is certainly not prime. You might say that this is a counterexample to the conjecture that $n^2 - n + 41$ is always prime. Since so many statements in mathematics are universal, making their negations existential, we can often prove that a statement is false (if it is) by providing a counterexample.

Example 4.2.10

Above we proved, "for all integers a and b, if $a + b$ is odd, then a is odd or b is odd." Is the converse true?

Solution

The converse is the statement, "for all integers a and b, if a is odd or b is odd, then a + b is odd." This is false! How do we prove it is false? We need to prove the negation of the converse.

Let's look at the symbols. The converse is

$$\forall a \forall b((O(a) \vee O(b)) \rightarrow O(a + b)).$$

We want to prove the negation:

$$\neg\forall a \forall b((O(a) \vee O(b)) \rightarrow O(a + b)).$$

Simplify using the rules from the previous sections:

$$\exists a \exists b((O(a) \vee O(b)) \wedge \neg O(a + b)).$$

As the negation passed by the quantifiers, they changed from $\forall$ to $\exists$. We then needed to take the negation of an implication, which is equivalent to asserting the if part and not the then part.

Now we know what to do. To prove that the converse is false we need to find two integers a and b so that a is odd or b is odd, but a + b is not odd (so even). That's easy: 1 and 3. (remember, "or" means one or the other or both). Both of these are odd, but 1 + 3 = 4 is not odd.

Proof by Cases

We could go on and on and on about different proof styles (we haven't even mentioned induction or combinatorial proofs here), but instead we will end with one final useful technique: proof by cases. The idea is to prove that P is true by proving that $Q \rightarrow P$ and $\neg Q \rightarrow P$ for some statement Q. So no matter what, whether or not Q is true, we know that P is true. In fact, we could generalize this. Suppose we want to prove P. We know that at least one of the statements $Q_1, Q_2, \ldots, Q_n$ is true. If we can show that $Q_1 \rightarrow P$ and $Q_2 \rightarrow P$ and so on all the way to $Q_n \rightarrow P$, then we can conclude P. The key thing is that we want to be sure that one of our cases (the Q_i's) must be true no matter what.

If that last paragraph was confusing, perhaps an example will make things better.

Example 4.2.11

Prove: For any integer n, the number $(n^3 - n)$ is even.

Solution

It is hard to know where to start this, because we don't know much of anything about n. We might be able to prove that $n^3 - n$ is even if we knew that n was even. In fact, we could probably prove that $n^3 - n$ was even if n was odd. But since n must either be even or odd, this will be enough. Here's the proof.

Proof. **We consider two cases:** if n is even or if n is odd.

Case 1: n is even. Then $n = 2k$ for some integer k. This give

$$n^3 - n = 8k^3 - 2k$$
$$= 2(4k^2 - k),$$

and since $4k^2$ k is an integer, this says that n3 n is even.

Case 2: n is odd. Then n = 2k + 1 for some integer k. This gives

$$n^3 - n = (2k+1)^3 - (2k+1)$$
$$= 8k^3 + 6k^2 + 6k + 1 - 2k - 1$$
$$= 2(4k^3 + 3k^2 + 2k),$$

and since $4k^3 + 3k^2 + 2k$ is an integer, we see that $n^3 - n$ is even again.

Since $n^3 - n$ is even in both exhaustive cases, we see that $n^3 - n$ is indeed always even.

Exercises

1. **Consider the statement "for all integers a and b, if** $a + b$ **is even, then** a **and** b **are even"**
   (a) Write the contrapositive of the statement.
   (b) Write the converse of the statement.
   (c) Write the negation of the statement.
   (d) Is the original statement true or false? Prove your answer.
   (e) Is the contrapositive of the original statement true or false? Prove your answer.
   (f) Is the converse of the original statement true or false? Prove your answer.
   (g) Is the negation of the original statement true or false? Prove your answer.
2. **For each of the statements below, say what method of proof you should use to prove them. Then say how the proof starts and how it ends. Bonus points for filling in the middle.**
   (a) There are no integers x and y such that x is a prime greater than 5 and $x = 6y + 3$.

(b) For all integers n, if n is a multiple of 3, then n can be written as the sum of consecutive integers.

(c) For all integers a and b, if $a^2 + b^2$ is odd, then a or b is odd.

3. **Consider the statement: for all integers n, if n is even then 8n is even.**

(a) Prove the statement. What sort of proof are you using?

(b) Is the converse true? Prove or disprove.

4. **The game TENZI comes with 40 six-sided dice (each numbered 1 to 6). Suppose you roll all 40 dice.**

(a) Prove that there will be at least seven dice that land on the same number.

(b) How many dice would you have to roll before you were guaranteed that some four of them would all match or all be different? Prove your answer.

5. **Prove that for all integers n, it is the case that n is even if and only if 3n is even. That is, prove both implications: if n is even, then 3n is even, and if 3n is even, then n is even.**

6. **Prove that $\sqrt{3}$ is irrational.**

7. **Consider the statement: for all integers a and b, if a is even and b is a multiple of 3, then ab is a multiple of 6.**

(a) Prove the statement. What sort of proof are you using?

(b) State the converse. Is it true? Prove or disprove.

8. **Prove the statement: For all integers n, if 5n is odd, then n is odd. Clearly state the style of proof you are using.**

9. **Prove the statement: For all integers a, b, and c, if $a^2 + b^2 = c^2$, then a or b is even.**

10. **Suppose that you would like to prove the following implication:**

For all numbers n, if n is prime then n is solitary.

Write out the beginning and end of the argument if you were to prove the statement,

(a) Directly

(b) By contrapositive

(c) By contradiction

You do not need to provide details for the proofs (since you do not know what solitary means). However, make sure that you provide the first few and last few lines of the proofs so that we can see that logical structure you would follow.

11. **Suppose you have a collection of 5-cent stamps and 8-cent stamps. We saw earlier that it is possible to make any amount of postage greater than 27 cents using combinations of both these types of stamps. But, let's ask some other questions:**

    (a) Prove that if you only use an even number of both types of stamps, the amount of postage you make must be even.

    (b) Suppose you made an even amount of postage. Prove that you used an even number of at least one of the types of stamps.

    (c) Suppose you made exactly 72 cents of postage. Prove that you used at least 6 of one type of stamp.

12. **Prove: $x = y$ if and only if $xy = \frac{(x+y)^2}{4}$. Note, you will need to prove two "directions" here: the "if" and the "only if" part.**

13. **Prove that log(7) is irrational.**

14. **Prove that there are no integer solutions to the equation $x^2 = 4y + 3$.**

15. **Prove that every prime number greater than 3 is either one more or one less than a multiple of 6.**

16. **Your "friend" has shown you a "proof" he wrote to show that 1 = 3. Here is the proof:**

Proof. I claim that 1 = 3. Of course we can do anything to one side of an equation as long as we also do it to the other side. So subtract 2 from both sides. This gives 1 = 1. Now square both sides, to get 1 = 1. And we all agree this is true.

What is going on here? Is your friend's argument valid? Is the argument a proof of the claim 1 = 3? Carefully explain using what we know about logic.

17. **A standard deck of 52 cards consists of 4 suites (hearts, diamonds, spades and clubs) each containing 13 different values (Ace, 2, 3, . . . , 10, J, Q, K). If you draw some number of cards at random you might or might not have a pair (two cards with the same value) or three cards all of the same suit. However, if you draw enough cards, you will be guaranteed to have these. For each of the following, find the smallest number of cards you would need to draw to be guaranteed having the specified cards. Prove your answers.**

    (a) Three of a kind (for example, three 7's).

    (b) A flush of five cards (for example, five hearts).

    (c) Three cards that are either all the same suit or all different suits.

18. **Suppose you are at a party with 19 of your closest friends (so including you, there are 20 people there). Explain why there must be least two people at the party who are friends with the same number of people at the party. Assume friendship is always reciprocated.**

19. **Your friend has given you his list of 115 best Doctor Who episodes (in order of greatness). It turns out that you have seen 60 of them. Prove that there are at least two episodes you have seen that are exactly four episodes apart on your friend's list.**

20. **Suppose you have an** $n \times n$ **chessboard but your dog has eaten one of the corner squares. Can you still cover the remaining squares with dominoes? What needs to be true about n? Give necessary and sufficient conditions (that is, say exactly which values of n work and which do not work). Prove your answers.**

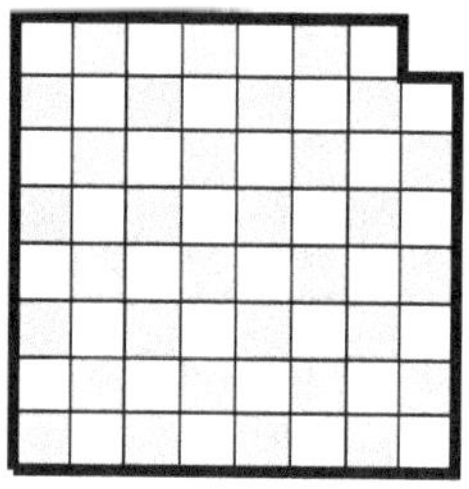

21. **What if your** $n \times n$ **chessboard is missing two opposite corners? Prove that no matter what n is, you will not be able to cover the remaining squares with dominoes.**

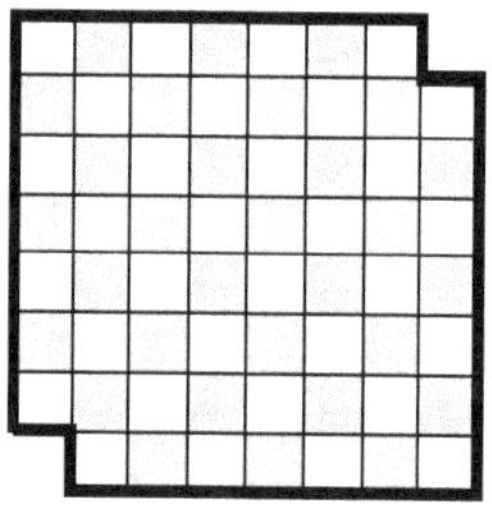

4.3 CONCISE SUMMARY

We have considered logic both as its own sub-discipline of mathematics, and as a means to help us better understand and write proofs. In either view, we noticed that mathematical statements have a particular logical form, and analyzing that form can help make sense of the statement.

At the most basic level, a statement might combine simpler statements using *logical connectives*. We often make use of variables, and *quantify* over those variables. How to resolve the truth or falsity of a statement based on these connectives and quantifiers is what logic is all about. From this, we can decide whether two statements are logically equivalent or if one or more statements (logically) imply another.

When writing proofs (in any area of mathematics) our goal is to explain why a mathematical statement is true. Thus it is vital that our argument implies the truth of the statement. To be sure of this, we first must know what it means for the statement to be true, as well as ensure that the statements that make up the proof correctly imply the conclusion. A firm understanding of logic is required to check whether a proof is correct.

There is, however, another reason that understanding logic can be helpful. Understanding the logical structure of a statement often gives clues as how to write a proof of the statement.

This is not to say that writing proofs is always straight forward.

Consider again the *Goldbach conjecture*:

Every even number greater than 2 can be written as the sum of two primes.

We are not going to try to prove the statement here, but we can at least say what a proof might look like, based on the logical form of the statement. Perhaps we should write the statement in an equivalent way which better highlights the quantifiers and connectives:

For all integers *n*, if *n* is even and greater than 2, then there exists integers *p* and *q* such that p and q are prime and *n* = p + q.

What would a direct proof look like? Since the statement starts with a universal quantifier, we would start by, ""Let n be an arbitrary integer." The rest of the statement is an implication. In a direct proof we assume the "if" part, so the next line would be, "Assume n is greater than 2 and is even." I have no idea what comes next, but eventually, we would need to find two prime numbers p and q (depending on n) and explain how we know that n = p + q.

Or maybe we try a proof by contradiction. To do this, we first assume

the negation of the statement we want to prove. What is the negation? From what we have studied we should be able to see that it is,

There is an integer n such that n is even and greater than 2, but for all integers p and q, either p or q is not prime or n ≠ p + q.

Could this statement be true? A proof by contradiction would start by assuming it was and eventually conclude with a contradiction, proving that our assumption of truth was incorrect. And if you can find such a contradiction, you will have proved the most famous open problem in mathematics. Good luck.

Chapter Review

1. **Complete a truth table for the statement $\neg P \rightarrow (Q \wedge R)$.**

2. **Suppose you know that the statement "if Peter is not tall, then Quincy is fat and Robert is skinny" is false. What, if anything, can you conclude about Peter and Robert if you know that Quincy is indeed fat? Explain (you may reference problem 3.3.1).**

3. **Are the statements $P \rightarrow (Q \vee R)$ and $(P \rightarrow Q) \vee (P \rightarrow R)$ logically equivalent? Explain your answer.**

4. **Is the following a valid deduction rule? Explain.**

$$\begin{array}{l} P \rightarrow Q \\ P \rightarrow R \\ \hline \therefore \quad P \rightarrow (Q \wedge R). \end{array}$$

5. **Write the negation, converse and contrapositive for each of the statements below.**

   (a) If the power goes off, then the food will spoil.

   (b) If the door is closed, then the light is off.

   (c) $\forall x(x < 1 \rightarrow x^2 < 1)$.

   (d) For all natural numbers n, if n is prime, then n is solitary.

   (e) For all functions f, if f is differentiable, then f is continuous.

   (f) For all integers a and b, if $a \cdot b$ is even, then a and b are even.

   (g) For every integer x and every integer y there is an integer n such that if x > 0 then $nx > y$.

   (h) For all real numbers x and y, if $xy = 0$ then $x = 0$ or $y = 0$.

   (i) For every student in Math 228, if they do not understand implications, then they will fail the exam.

6. **Consider the statement: for all integers n, if n is even and $n \leq 7$ then n is negative or $n \in \{0, 2, 4, 6\}$.**

   (a) Is the statement true? Explain why.

   (b) Write the negation of the statement. Is it true? Explain.

   (c) State the contrapositive of the statement. Is it true? Explain.

   (d) State the converse of the statement. Is it true? Explain.

7. **Consider the statement: $\forall x(\forall y(x + y = y) \rightarrow \forall z(x \cdot z = 0))$.**

   (a) Explain what the statement says in words. Is this statement true? Be sure to state what you are taking the universe of discourse to be.

   (b) Write the converse of the statement, both in words and in symbols. Is the converse true?

   (c) Write the contrapositive of the statement, both in words and in symbols. Is the contrapositive true?

(d) Write the negation of the statement, both in words and in symbols. Is the negation true?

8. Simplify the following.

(a) $\neg(\neg(P \wedge \neg Q) \rightarrow \neg(\neg R \vee \neg(P \rightarrow R)))$.

(b) $\neg\exists x \neg\forall y \neg\exists z(z = x + y \rightarrow \exists w(x - y = w))$.

9. Consider the statement: for all integers n, if n is odd, then 7n is odd.

(a) Prove the statement. What sort of proof are you using?

(b) Prove the converse. What sort of proof are you using?

10. Suppose you break your piggy bank and scoop up a handful of 22 coins (pennies, nickels, dimes and quarters).

(a) Prove that you must have at least 6 coins of a single denomination.

(b) Suppose you have an odd number of pennies. Prove that you must have an odd number of at least one of the other types of coins.

(c) How many coins would you need to scoop up to be sure that you either had 4 coins that were all the same or 4 coins that were all different? Prove your answer.

11. You come across four trolls playing bridge. They declare:

Troll 1: All trolls here see at least one knave.

Troll 2: I see at least one troll that sees only knaves.

Troll 3: Some trolls are scared of goats.

Troll 4: All trolls are scared of goats.

Are there any trolls that are not scared of goats? Recall, of course, that all trolls are either knights (who always tell the truth) or knaves (who always lie).

FOOTNOTES

1. This is not to say that looking at examples is a waste of time. Doing so will often give you an idea of how to write a proof. But the examples do not belong in the proof.

CHAPTER-5

GRAPH THEORY

INTRODUCTION

Graph theory is a branch of mathematics that explores the properties and relationships of graphs. A graph in this context is not a plotted curve but rather a collection of vertices (nodes) connected by edges (lines). This abstract representation allows us to model and analyze a wide range of phenomena from networks of social interactions to the structure of molecules, from transportation systems to the internet. Graph theory provides a powerful framework for understanding connectivity, paths, cycles, and other structural properties within these networks. Its applications span disciplines such as computer science, biology, economics, and sociology, making it an indispensable tool for solving real-world problems and advancing our understanding of complex systems. This introduction delves into the fundamental concepts of graph theory, its historical development, key applications, and its significance in various fields of study.

Investigate!

In the time of Euler, in the town of Königsberg in Prussia, there was a river containing two islands. The islands were connected to the banks of the river by seven bridges (as seen below). The bridges were very beautiful, and on their days off, townspeople would spend time walking over the bridges. As time passed, a question arose: was it possible to plan a walk so that you cross each bridge once and only once? Euler was able to answer this question. Are you?

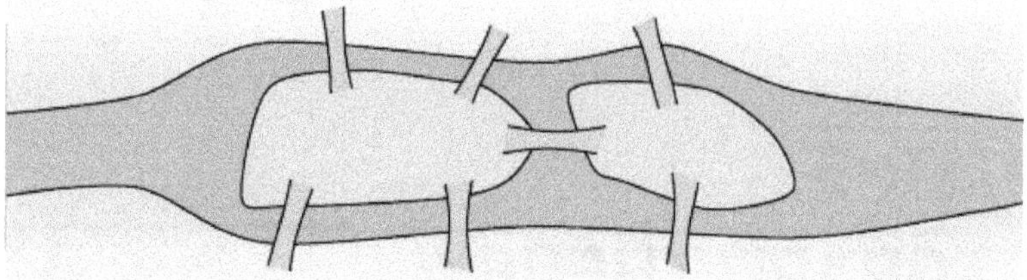

Attempt the above activity before proceeding

Graph Theory is a relatively new area of mathematics, first studied by the super famous mathematician Leonhard Euler in 1735. Since then it has blossomed in to a powerful tool used in nearly every branch of science and is currently an active area of mathematics research.

The problem above, known as the *Seven Bridges of Königsberg,* is the problem that originally inspired graph theory. Consider a "different" problem: Below is a drawing of four dots connected by some lines. Is it possible to trace over each line once and only once (without lifting up your pencil, starting and ending on a dot)?

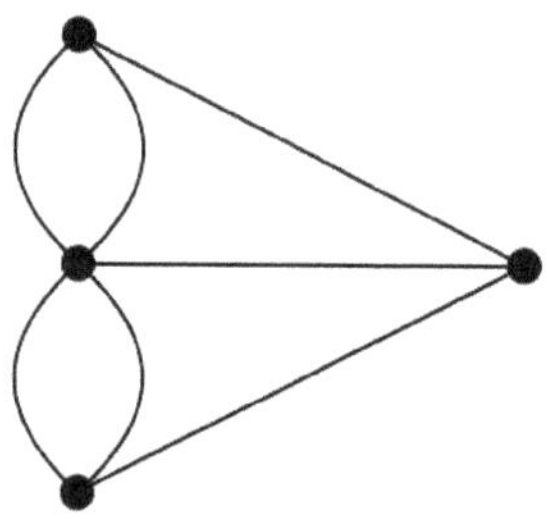

There is an obvious connection between these two problems. Any path in the dot and line drawing corresponds exactly to a path over the bridges of Königsberg.

Pictures like the dot and line drawing are called **graphs.** Graphs are made up of a collection of dots called **vertices** and lines connecting those dots called **edges.** When two vertices are connected by an edge, we say they are **adjacent.** The nice thing about looking at graphs instead of pictures of rivers, islands and bridges is that we now have a mathematical object to study. We have distilled the "important" parts of the bridge picture for the purposes of the problem. It does not matter how big the islands are, what the bridges are made out of, if the river contains alligators, etc. All that matters is which land masses are connected to which other land masses, and how many times.

We will return to the question of finding paths through graphs later. But first, here are a few other situations you can represent with graphs:

Example 5.0.1

Al, Bob, Cam, Dan, and Euler are all members of the social networking website Facebook. The site allows members to be "friends" with each other. It turns out that Al and Cam are friends, as are Bob and Dan. Euler is friends with everyone. Represent this situation with a graph.

Solution

Each person will be represented by a vertex and each friendship will be represented by an edge. That is, two vertices will be adjacent (there will be an edge between them) if and only if the people represented by those vertices are friends.

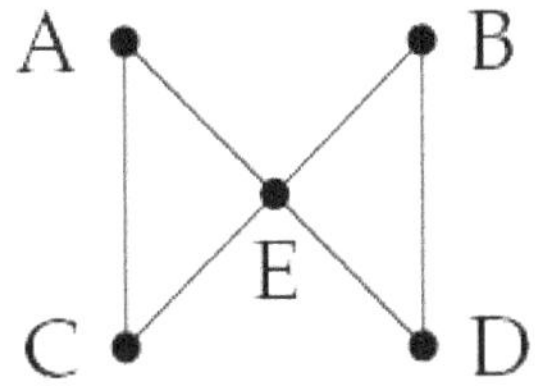

Example 5.0.2

Each of three houses must be connected to each of three utilities. Is it possible to do this without any of the utility lines crossing?

Solution

We will answer this question later. For now, notice how we would ask this question in the context of graph theory. We are really asking whether it is possible to redraw the graph below without any edges crossing (except at vertices). Think of the top row as the houses, bottom row as the utilities.

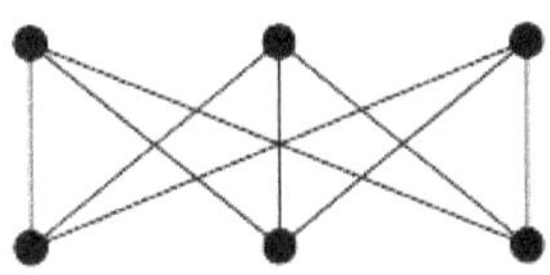

5.1 DEFINITIONS

Investigate!

Which (if any) of the graphs below are the same?

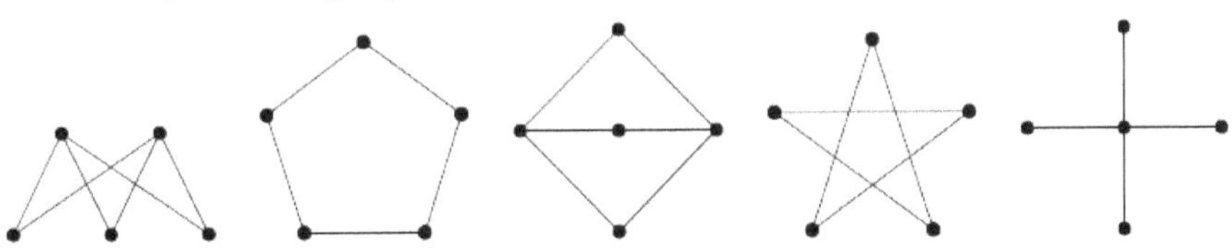

The graphs above are unlabeled. Usually we think of a graph as having a specific set of vertices. Which (if any) of the graphs below are the same?

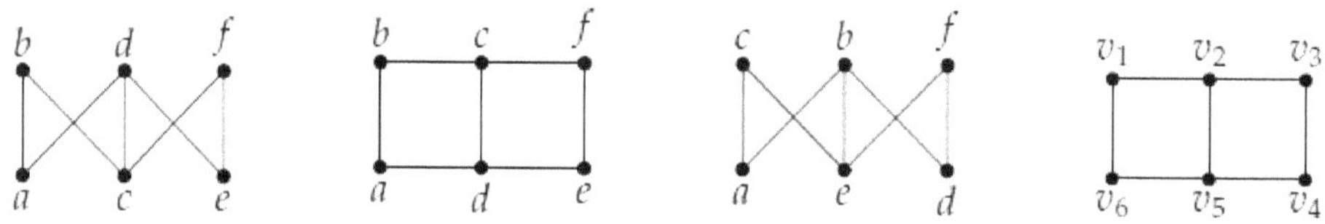

Actually, all the graphs we have seen above are just drawings of graphs. A graph is really an abstract mathematical object consisting of two sets V and E where E is a set of 2-element subsets of V. Are the graphs below the same or different?

Graph 1:

$V = \{a, b, c, d, e\}$,

$E = \{\{a, b\}, \{a, c\}, \{a, d\}, \{a, e\}, \{b, c\}, \{d, e\}\}$.

Graph 2:

$V = \{v_1, v_2, v_3, v_4, v_5\}$,

$E = \{\{v_1, v_3\}, \{v_1, v_5\}, \{v_2, v_4\}, \{v_2, v_5\}, \{v_3, v_5\}, \{v_4, v_5\}\}$.

Attempt the above activity before proceeding

Before we start studying graphs, we need to agree upon what a graph is. While we almost always think of graphs as pictures (dots connected by lines) this is fairly ambiguous. Do the lines need to be straight? Does it matter how long the lines are or how large the dots are? Can there be two lines connecting the same pair of dots? Can one line connect three dots? The way we avoid ambiguities in mathematics is to provide concrete and rigorous *definitions.* Crafting good definitions is not easy, but it is incredibly important. The definition is the agreed upon starting point from which all truths in mathematics proceed. Is there a graph with no edges? We have to look at the definition to see if this is possible.

We want our definition to be precise and unambiguous, but it also must agree with our intuition for the objects we are studying. It needs to be useful: we could define a graph to be a six legged mammal, but that would not let us solve any problems about bridges. Instead, here is the (now) standard definition of a graph.

Graph Definition

A **graph** is an ordered pair $G = (V, E)$ consisting of a nonempty set V (called the **vertices**) and a set E (called the **edges**) of two-element subsets of V.

Strange. Nowhere in the definition is there talk of dots or lines. From the definition, a graph could be

$$(\{a, b, c, d\}, \{\{a, b\}, \{a, c\}, \{b, c\}, \{b, d\}, \{c, d\}\}).$$

Here we have a graph with four vertices (the letters a, b, c, d) and five edges (the pairs $\{a, b\}, \{a, c\}, \{b, c\}, \{b, d\}, \{c, d\}$)).

Looking at sets and sets of 2-element sets is difficult to process. That is why we often draw a representation of these sets. We put a dot down for each vertex, and connect two dots with a line precisely when those two vertices are one of the 2-element subsets in our set of edges. Thus one way to draw the graph described above is this:

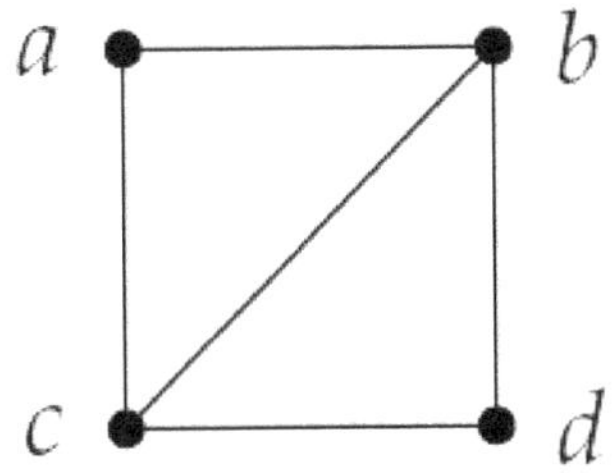

However we could also have drawn the graph differently. For example either of these:

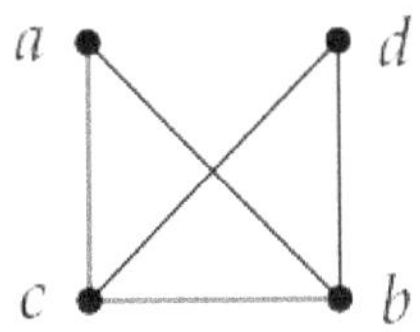

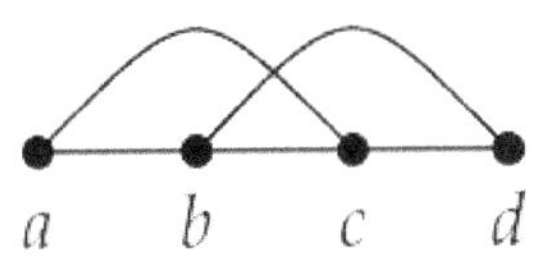

We should be careful about what it means for two graphs to be "the same." Actually, given our definition, this is easy: Are the vertex sets equal? Are the edge sets equal? We know what it means for sets to be equal, and graphs are nothing but a pair of two special sorts of sets.

Example 5.1.1

Are the graphs below equal?

$G_1 = (\{a, b, c\}, \{\{a, b\}, \{b, c\}\})$; $G_2 = (\{a, b, c\}, \{\{a, c\}, \{c, b\}\})$.

Solution

No. Here the vertex sets of each graph are equal, which is a good start. Also, both graphs have two edges. In the first graph, we have edges $\{a, b\}$ and $\{b, c\}$, while in the second graph we have edges $\{a, c\}$ and $\{c, b\}$. Now we do have $\{b, c\} = \{c, b\}$, so that is not the problem. The issue is that $\{a, b\} \neq \{a, c\}$. Since the edge sets of the two graphs are not equal (as sets), the graphs are not equal (as graphs).

Even if two graphs are not *equal*, they might be basically the same. The graphs in the previous example could be drawn like this:

Graphs that are basically the same (but perhaps not equal) are called **isomorphic.** We will give a precise definition of this term after a quick example:

Example 5.1.2

Consider the graphs:

- $G_1 = (V_1, E_1)$ where $V_1 = \{a, b, c\}$ and $E_1 = \{\{a, b\}, \{a, c\}, \{b, c\}\}$;
- $G_2 = (V_2, E_2)$ where $V_2 = \{u, v, w\}$ and $E_2 = \{\{u, v\}, \{u, w\}, \{v, w\}\}$.

Are these graphs the same?

Solution

The two graphs are NOT equal. It is enough to notice that $V_1 \neq V_2$ since a $\in$ V_1 but a $\notin$ <1 V_2. However, both of these graphs consist of three vertices with edges connecting every pair of vertices. We can draw them as follows:

Clearly we want to say these graphs are basically the same, so while they are not equal, they will be *isomorphic*. We can rename the vertices of one graph and get the second graph as the result.

Intuitively, graphs are **isomorphic** if they are basically the same, or better yet, if they are the same except for the names of the vertices. To make the concept of renaming vertices precise, we give the following definitions:

Isomorphic Graphs

An **isomorphism** between two graphs G_1 and G_2 is a bijection $f: V_1 \rightarrow V_2$ between the vertices of the graphs such that $\{a, b\}$ is an edge in G_1 if and only if $\{f(a), f(b)\}$ is an edge in G_2.

Two graphs are **isomorphic** if there is an isomorphism between them. In this case we write $G_1 \cong G_2$.

An isomorphism is simply a function which renames the vertices. It must be a bijection so every vertex gets a new name. These newly named vertices must be connected by edges precisely when they were connected by edges with their old names.

Example 5.1.3

Decide whether the graphs $G_1 = (V_1, E_1)$ and $G_2 = (V_2, E_2)$ are equal or isomorphic.

- $V_1 = \{a, b, c, d\}$, $E_1 = \{\{a, b\}, \{a, c\}, \{a, d\}, \{c, d\}\}$
- $V_2 = \{a, b, c, d\}$, $E_2 = \{\{a, b\}, \{a, c\}, \{b, c\}, \{c, d\}\}$

Solution

The graphs are NOT equal, since $\{a, d\} \in E_1$ but $\{a, d\} \notin E_2$. However, since both graphs contain the same number of vertices and same number of edges, they *might* be isomorphic (this is not enough in most cases, but it is a good start).

We can try to build an isomorphism. How about we say $f(a) = b$, $f(b) = c$, $f(c) = d$ and $f(d) = a$. This is definitely a bijection, but to make sure that the function is an isomorphism, we must make sure it respects the edge relation. In G_1, vertices a and b are connected by an edge. In G_2, $f(a) = b$ and $f(b) = c$ are connected by an edge. So far, so good, but we must check the other three edges. The edge $\{a, c\}$ in G_1 corresponds to $\{f(a), f(c)\} = \{b, d\}$, but here we have a problem. There is no edge between b and d in G_2. Thus f is NOT an isomorphism.

Not all hope is lost, however. Just because f is not an isomorphism does not mean that there is no isomorphism at all. We can try again. At this point it might be helpful to draw the graphs to see how they should match up.

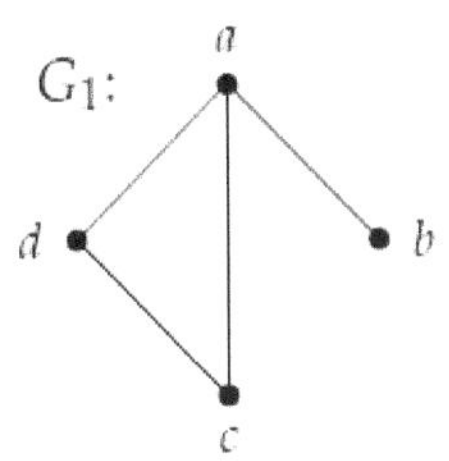

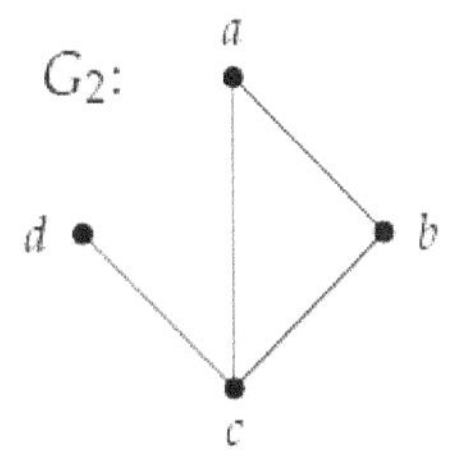

Alternatively, notice that in G_1, the vertex a is adjacent to every other vertex. In G_2, there is also a vertex with this property: c. So build the bijection $g: V_1 \to V_2$ by defining $g(a) = c$ to start with.

where should we send b? In G_1, the vertex b is only adjacent to vertex a. There is exactly one vertex like this in G_2, namely d. So let $g(b) = d$. As for the last two, in this example, we have a free choice: let $g(c) = b$ and $g(d) = a$ (switching these would be fine as well).

We should check that this really is an isomorphism. It is definitely a bijection. We must make sure that the edges are respected. The four edges in G_1 are

$$\{a, b\}, \{a, c\}, \{a, d\}, \{c, d\}.$$

Under the proposed isomorphism these become

$$\{g(a), g(b)\}, \{g(a), g(c)\}, \{g(a), g(d)\}, \{g(c), g(d)\}$$

$$\{c, d\}, \{c, b\}, \{c, a\}, \{b, a\}, X$$

which are precisely the edges in G_2. Thus g is an isomorphism, so $G_1 \cong G_2$

Sometimes we will talk about a graph with a special name (like K_n or the Peterson graph) or perhaps draw a graph without any labels. In this case we are really referring to all graphs isomorphic to any copy of that particular graph. A collection of isomorphic graphs is often called an **isomorphism class.**

There are other relationships between graphs that we care about, other than equality and being isomorphic. For example, compare the following pair of graphs:

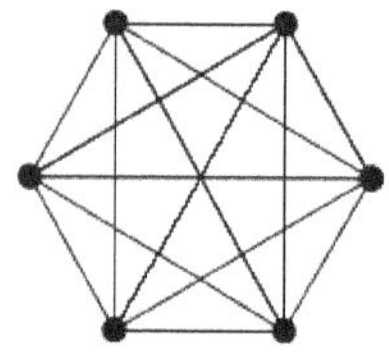

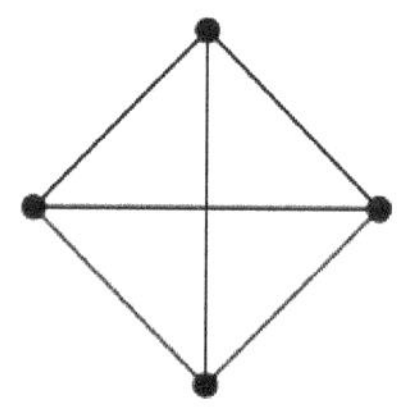

These are definitely not isomorphic, but notice that the graph on the right looks like it might be part of the graph on the left, especially if we draw it like this:

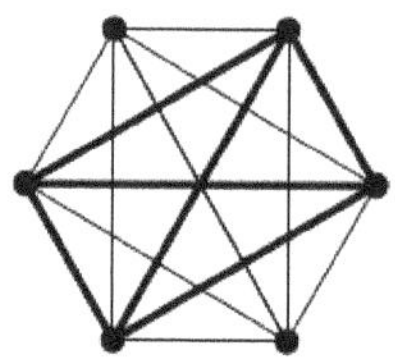

We would like to say that the smaller graph is a subgraph of the larger.

We should give a careful definition of this. In fact, there are two reasonable notions for what a subgraph should mean.

Subgraphs

We say that $G' = (V', E')$ is a **subgraph** of $G = (V, E)$, and write $G' \subseteq G$, provided $V' \subseteq V$ and $E' \subseteq E$.

We say that $G' = (V', E')$ is an **induced subgraph** of $G = (V, E)$ provided $V' \subseteq V$ and every edge in E whose vertices are still in V' is also an edge in E'.

Notice that every induced subgraph is also an ordinary subgraph, but not conversely. Think of a subgraph as the result of deleting some vertices and edges from the larger graph. For the subgraph to be an induced subgraph, we can still delete vertices, but now we only delete those edges that included the deleted vertices.

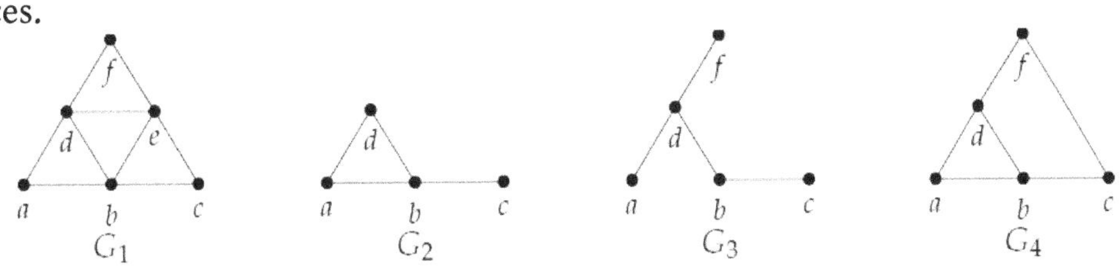

Here both G_2 and G_3 are subgraphs of G_1. But only G_2 is an induced subgraph. Every edge in G_1 that connects vertices in G_2 is also an edge in G_2. In G_3, the edge $\{a, b\}$ is in E_1 but not E_3, even though vertices a and b are in V_3.

The graph G_4 is NOT a subgraph of G_1, even though it looks like all we did is remove vertex e. The reason is that in E_4 we have the edge $\{c, f\}$ but this is not an element of E_1, so we don't have the required $E_4 \subseteq E_1$.

Back to some basic graph theory definitions. Notice that all the graphs we have drawn above have the property that no pair of vertices is connected more than once, and no vertex is connected to itself. Graphs like these are sometimes called **simple**, although we will just call them *graphs*. This is because our definition for a graph says that the edges form a set of 2-element subsets of the vertices. Remember that it doesn't make sense to say a set contains an element more than once. So no pair of vertices can be connected by an edge more than once. Also, since each edge must be a set containing two vertices, we cannot have a single vertex connected to itself by an edge.

That said, there are times we want to consider double (or more) edges and single edge loops. For example, the "graph" we drew for the Bridges of Königsberg problem had double edges because there really are two bridges connecting a particular island to the near shore. We will call these objects **multigraphs.** This is a good name: a *multiset* is a set in which we are allowed to include a single element multiple times.

The graphs above are also **connected**: you can get from any vertex to any other vertex by following some path of edges. A graph that is not connected can be thought of as two separate graphs drawn close together. For example, the following graph is NOT connected because there is no path from a to b:

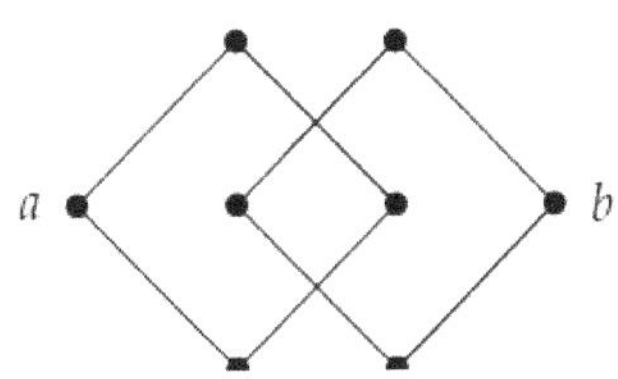

Vertices in a graph do not always have edges between them. If we add all possible edges, then the resulting graph is called **complete**. That is, a graph is complete if every pair of vertices is connected by an edge. Since a graph is determined completely by which vertices are adjacent to which other vertices, there is only one complete graph with a given number of vertices. We give these a special name: K_n is the complete graph on n vertices.

Each vertex in K_n is adjacent to $n-1$ other vertices. We call the number of edges emanating from a given vertex the **degree** of that vertex. So every vertex in K_n has degree $n-1$. How many edges does K_n have? One might think the answer should be $n(n-1)$, since we count $n-1$ edges n times (once for each vertex). However, each edge is incident to 2 vertices, so we counted every edge exactly twice. Thus there are $n(n-1)/2$ edges in K_n. Alternatively, we can say there are $\binom{n}{2}$ edges, since to draw an edge we must choose 2 of the n vertices.

In general, if we know the degrees of all the vertices in a graph, we can find the number of edges. The sum of the degrees of all vertices will always be twice the number of edges, since each edge adds to the degree of two vertices. Notice this means that the sum of the degrees of all vertices in any graph must be even!

This is our first example of a general result about all graphs. It seems innocent enough, but we will use it to prove all sorts of other statements. So let's give it a name and state it formally.

Lemma 5.1.5 Handshake Lemma. *In any graph, the sum of the degrees of vertices in the graph is always twice the number of edges.*

The handshake lemma[2] is sometimes called the degree sum formula, and can be written symbolically as

$$\sum_{v \in V} d(v) = 2e.$$

Here we are using the notation $d(v)$ for the degree of the vertex v.

One use for the lemma is to actually find the number of edges in a graph. To do this, you must be given the **degree sequence** for the graph (or be able to find it from other information). This is a list of every degree of every vertex in the graph, generally written in non-increasing order.

Example 5.1.6

How many vertices and edges must a graph have if its degree sequence is (4, 4, 3, 3, 3, 2, 1)?

Solution

The number of vertices is easy to find: it is the number of degrees in the sequence: 7. To find the number of edges, we compute the degree sum:

$$4 + 4 + 3 + 3 + 3 + 2 + 1 = 20,$$

so the number of edges is half this: 10.

The handshake lemma also tells us what is not possible.

Example 5.1.7

At a recent math seminar, 9 mathematicians greeted each other by shaking hands. Is it possible that each mathematician shook hands with exactly 7 people at the seminar?

Solution

It seems like this should be possible. Each mathematician chooses one person to not shake hands with. But this cannot happen. We are asking whether a graph with 9 vertices can have each vertex have degree 7. If such a graph existed, the sum of the degrees of the vertices would be $9 \cdot 7 = 63$. This would be twice the number of edges (handshakes) resulting in a graph with 31.5 edges. That is impossible. Thus at least one (in fact an odd number) of the mathematicians must have shaken hands with an *even* number of people at the seminar.

We can generalize the previous example to get the following proposition.[3]

Proposition 5.1.8 *In any graph, the number of vertices with odd degree must be even.*

Proof. Suppose there were a graph with an odd number of vertices with odd degree. Then the sum of the degrees in the graph would be odd, which is impossible, by the handshake lemma.

We will consider further applications of the handshake lemma in the exercises.

One final definition: we say a graph is **bipartite** if the vertices can be divided into two sets, *A* and *B*, with no two vertices in A adjacent and no two vertices in B adjacent. The vertices in A can be adjacent to some or all of the vertices in B. If each vertex in A is adjacent to all the vertices in B, then the graph is a **complete bipartite graph**, and gets a special name: $K_{m,n}$, where $|A| = m$ and $|B| = n$. The graph in the houses and utilities puzzle is $K_{3,3}$.

Named Graphs

Some graphs are used more than others, and get special names.

- K_n The complete graph on n vertices.
- $K_{m,n}$ The complete bipartite graph with sets of m and n vertices.
- C_n The cycle on *n* vertices, just one big loop.

- P_n The path on $n + 1$ vertices (so n edges), just one long path.

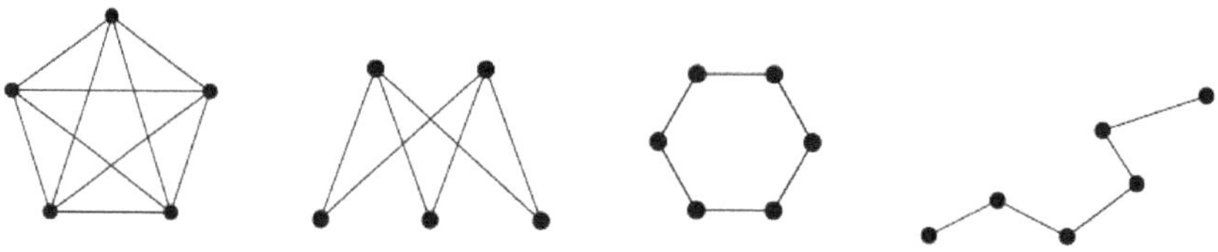

K_5 $K_{2,3}$ C_6 P_5

i

Graph Theory Definitions

There are a lot of definitions to keep track of in graph theory. Here is a glossary of the terms we have already used and will soon encounter.

Graph

A collection of **vertices**, some of which are connected by **edges**. More precisely, a pair of sets V and E where V is a set of vertices and E is a set of 2-element subsets of V.

Adjacent

Two vertices are **adjacent** if they are connected by an edge. Two edges are **adjacent** if they share a vertex.

Bipartite graph

A graph for which it is possible to divide the vertices into two disjoint sets such that there are no edges between any two vertices in the same set.

Complete bipartite graph

A bipartite graph for which every vertex in the first set is adjacent to every vertex in the second set.

Complete graph

A graph in which every pair of vertices is adjacent.

Connected

A graph is **connected** if there is a path from any vertex to any other vertex.

Chromatic number

The minimum number of colors required in a proper vertex coloring of the graph.

Cycle

A path (see below) that starts and stops at the same vertex, but contains no other repeated vertices.

Degree of a vertex

The number of edges incident to a vertex.

Euler path

A walk which uses each edge exactly once.

Euler circuit

An Euler path which starts and stops at the same vertex.

Multigraph

A **multigraph** is just like a graph but can contain multiple edges between two vertices as well as single edge loops (that is an edge from a vertex to itself).

Path

A **path** is a walk that doesn't repeat any vertices (or edges) except perhaps the first and last. If a path starts and ends at the same vertex, it is called a **cycle.**

Planar

A graph which can be drawn (in the plane) without any edges crossing.

Subgraph

We say that H is a **subgraph** of G if every vertex and edge of H is also a vertex or edge of G. We say H is an **induced** subgraph of G if every vertex of H is a vertex of G and each pair of vertices in H are adjacent in H if and only if they are adjacent in G.

Tree

A connected graph with no cycles. (If we remove the requirement that the graph is connected, the graph is called a **forest.**) The vertices in a tree with degree 1 are called **leaves.**

Vertex coloring

An assignment of colors to each of the vertices of a graph. A vertex coloring is **proper** if adjacent vertices are always colored differently.

Walk

A sequence of vertices such that consecutive vertices (in the se- quence) are adjacent (in the graph). A walk in which no edge is repeated is called a **trail,** and a trail in which no vertex is repeated (except possibly the first and last) is called a **path.**

Exercises

1. If 10 people each shake hands with each other, how many handshakes took place? What does this question have to do with graph theory?
2. Among a group of 5 people, is it possible for everyone to be friends with exactly 2 of the people in the group? What about 3 of the people in the group?
3. Is it possible for two different (non-isomorphic) graphs to have the same number of vertices and the same number of edges? What if the degrees of the vertices in the two graphs are the same (so both graphs have vertices with degrees 1, 2, 2, 3, and 4, for example)? Draw two such graphs or explain why not.
4. Are the two graphs below equal? Are they isomorphic? If they are isomorphic, give the isomorphism. If not, explain.

Graph 1:

$$V = \{a, b, c, d, e\}, E = \{\{a, b\}, \{a, c\}, \{a, e\}, \{b, d\}, \{b, e\}, \{c, d\}\}.$$

Graph 2:

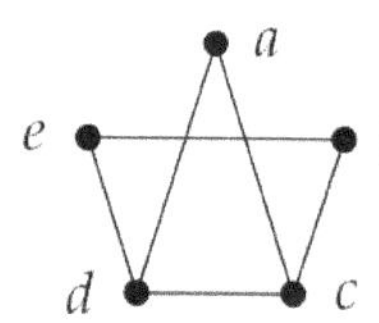

5. Consider the following two graphs:

G_1 $V_1 = \{a, b, c, d, e, f, g\}$

$E_1 = \{\{a, b\}, \{a, d\}, \{b, c\}, \{b, d\}, \{b, e\}, \{b, f\}, \{c, g\}, \{d, e\}, \{e, f\}, \{f, g\}\}.$

G_2 $V_2 = \{v_1, v_2, v_3, v_4, v_5, v_6, v_7\},$

$E_2 = \{\{v_1, v_4\}, \{v_1, v_5\}, \{v_1, v_7\}, \{v_2, v_3\}, \{v_2, v_6\}, \{v_3, v_5\}, \{v_3, v_7\}, \{v_4, v_5\}, \{v_5, v_6\}, \{v_5, v_7\}\}$

(a) Let $f: G_1 \to G_2$ be a function that takes the vertices of Graph 1 to vertices of Graph 2. The function is given by the following table:

x	a	b	c	d	e	f	g
$f(x)$	v_4	v_5	v_1	v_6	v_2	v_3	v_7

Does f define an isomorphism between Graph 1 and Graph 2?

(b) Define a new function g (with $g \neq f$) that defines an isomorphism between Graph 1 and Graph 2.

(c) Is the graph pictured below isomorphic to Graph 1 and Graph 2? Explain.

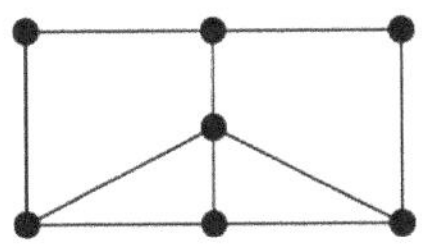

6. **What is the largest number of edges possible in a graph with 10 vertices? What is the largest number of edges possible in a bipartite graph with 10 vertices? What is the largest number of edges possible in a tree with 10 vertices?**

7. **Which of the graphs below are bipartite? Justify your answers.**

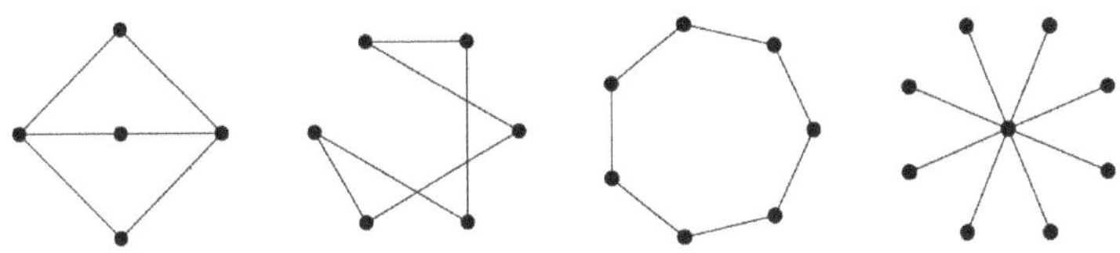

8. **For which $n \geq 3$ is the graph C_n bipartite?**

9. **For each of the following, try to give two *different* unlabeled graphs with the given properties, or explain why doing so is impossible.**

   (a) wo different trees with the same number of vertices and the same number of edges. A tree is a connected graph with no cycles.

   (b) Two different graphs with 8 vertices all of degree 2.

   (c) Two different graphs with 5 vertices all of degree 4.

   (d) Two different graphs with 5 vertices all of degree 3.

10. **Decide whether the statements below about subgraphs are true or false. For those that are true, briefly explain why (1 or 2 sentences). For any that are false, give a counterexample.**

    (a) Any subgraph of a complete graph is also complete.

    (b) Any *induced* subgraph of a complete graph is also complete.

    (c) Any subgraph of a bipartite graph is bipartite.

    (c) Any subgraph of a tree is a tree.

11. **Let $k_1, k_2, \ldots, k_j$ be a list of positive integers that sum to n (i.e., $\sum_{i=1}^{j} k_i = n$). Use two graphs containing n vertices to explain why**

$$\sum_{i=1}^{j} \binom{k_i}{2} \leq \binom{n}{2}.$$

12. **We often define graph theory concepts using set theory. For example, given a graph $G = (V, E)$ and a vertex $v \in V$, we define**

$$N(v) = \{u \in V : \{v, u\} \in E\}.$$

We define $N[v] = N[v] \cup \{v\}$. The goal of this problem is to figure out what all this means.

(a) Let G be the graph with $V = \{a, b, c, d, e, f\}$ and $E = \{\{a, b\}, \{a, e\}, \{b, c\}, \{b, e\}, \{c, d\}, \{c, f\}, \{d, f\}, \{e, f\}\}$. Find $N(a)$, $N[a]$, $N(c)$, and $N[c]$.

(b) What is the largest and smallest possible values for $|N(v)|$ and $|N[v]|$ for the graph in part (a)? Explain.

(c) Give an example of a graph $G = (V, E)$ (probably different than the one above) for which $N[v] = V$ for some vertex $v \in V$. Is there a graph for which $N[v] = V$ for all $v \in V$? Explain.

(d) Give an example of a graph $G = (V, E)$ for which $N(v) = \emptyset$ for some $v \in V$. Is there an example of such a graph for which $N[u] = V$ for some other $u \in V$ as well? Explain.

(e) Describe in words what $N(v)$ and $N[v]$ mean in general.

13. **A graph is a way of representing the relationships between elements in a set: an edge between the vertices x and y tells us that x is related to y (which we can write as x ~ y). Not all sorts of relationships can be represented by a graph though. For each relationship described below, either draw the graph or explain why the relationship cannot be represented by a graph.**

(a) The set V = {1, 2, . . . , 9} and the relationship $x \sim y$ when $x - y$ is a non-zero multiple of 3.

(b) The set V = {1, 2, . . . , 9} and the relationship $x \sim y$ when y is a multiple of x.

(c) The set V = {1, 2, . . . , 9} and the relationship x ~ y when $0 < |x - y| < 3$.

14. **Consider graphs with n vertices. Remember, graphs do not need to be *connected.***

(a) How many edges must the graph have to guarantee at least one vertex has degree two or more? Prove your answer.

(b) How many edges must the graph have to guarantee all vertices have degree two or more? Prove your answer.

15 **.Prove that any graph with at least two vertices must have two vertices of the same degree.**

16. **Suppose G is a connected graph with $n > 1$ vertices and $n - 1$ edges. Prove that G has a vertex of degree 1.**

5.2 TREES

Investigate!

Consider the graph drawn below.

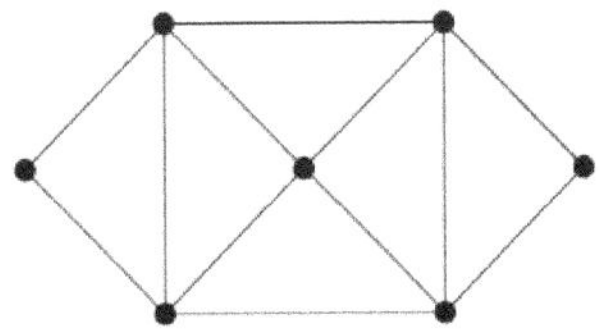

- Find a subgraph with the smallest number of edges that is still connected and contains all the vertices.
- Find a subgraph with the largest number of edges that doesn't contain any cycles.
- What do you notice about the number of edges in your examples above? Is this a coincidence?

Attempt the above activity before proceeding

One very useful and common approach to studying graph theory is to restrict your focus to graphs of a particular kind. For example, you could try to really understand just complete graphs or just bipartite graphs, instead of trying to understand all graphs in general. That is what we are going to do now, looking at trees. Hopefully by the end of this section we will have a better understanding of this class of graph, and also understand why it is important enough to warrant its own section.

Definition of a Tree

- A **tree** is a connected graph containing no cycles.[4]
- A **forest** is a graph containing no cycles. Note that this means that a connected forest is a tree.

Does the definition above agree with your intuition for what graphs we should call trees? Try thinking of examples of trees and make sure they satisfy the definition. One thing to keep in mind is that while the trees we study in graph theory are related to trees you might see in other subjects, the correspondence is not exact. For instance, in anthropology, you might study family trees, like the one below,

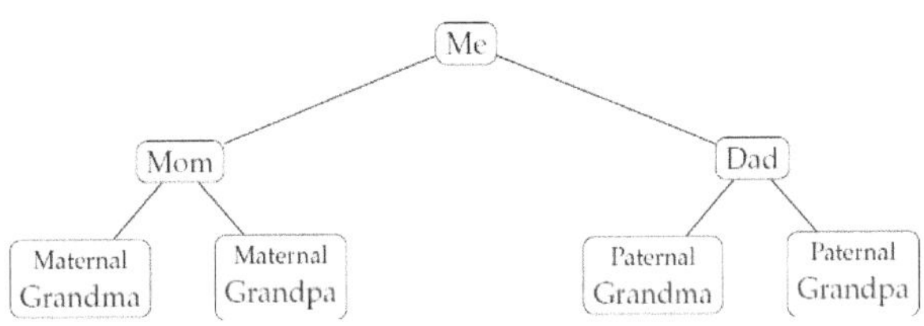

So far so good, but while your grandparents are (probably) not blood- relatives, if we go back far enough, it is likely that they did have some common ancestor. If you trace the tree back from you to that common ancestor, then down through your other grandparent, you would have a cycle, and thus the graph would not be a tree.

You might also have seen something called a *decision tree* (such as the algorithm for deciding whether a series converges or diverges). Sometimes these too contain cycles, as the decision for one node might lead you back to a previous step.

Both the examples of trees above also have another feature worth mentioning: there is a clear order to the vertices in the tree. In general, there is no reason for a tree to have this added structure, although we can impose such a structure by considering rooted trees, where we simply designate one vertex as the root. We will consider such trees in more detail later in this section.

Properties of Trees

We wish to really understand trees. This means we should discover properties of trees; what makes them special and what is special about them.

A tree is a connected graph with no cycles. Is there anything else we can say? It would be nice to have other equivalent conditions for a graph to be a tree. That is, we would like to know whether there are any graph theoretic properties that all trees have, and perhaps even that only trees have.

To get a feel for the sorts of things we can say, we will consider three *propositions* about trees. These will also illustrate important proof techniques that apply to graphs in general, and happen to be a little easier for trees.

Our first proposition gives an alternate definition for a tree. That is, it gives necessary and sufficient conditions for a graph to be a tree.

Proposition 5.2.1 *A graph T is a tree if and only if between every pair of distinct vertices of T there is a unique path.*

Proof. This is an "if and only if" statement, so we must prove two implications. We start by proving that if T is a tree, then between every pair of distinct vertices there is a unique path.

Assume T is a tree, and let u and v be distinct vertices (if T only has one vertex, then the conclusion is satisfied automatically). We must show two things to show that there is a unique path between u and v: that there is a path, and that there is not more than one path. The first of these is automatic, since T is a tree, it is connected, so there is a path between any pair of vertices.

To show the path is unique, we suppose there are two paths between u and v, and get a contradiction. The two paths might start out the same, but since they are different, there is some first vertex u' after which the two paths diverge. However, since the two paths both end and v, there is some first vertex after *u'* that they have in common, call it *v'*. Now consider the two paths from *u'* to *v'*. Taken together, these form a cycle, which contradicts our assumption that T is a tree.

Now we consider the converse: if between every pair of distinct vertices of T there is a unique path, then T is a tree. So assume the hypothesis: between every pair of distinct vertices of T there is a unique path. To prove that T is a tree, we must show it is connected and contains no cycles.

The first half of this is easy: T is connected, because there is a path between every pair of vertices. To show that T has no cycles, we assume it does, for the sake of contradiction. Let u and v be two distinct vertices in a cycle of T. Since we can get from *u* to *v* by going clockwise or counterclockwise around the cycle, there are two paths from *u* and *v*, contradicting our assumption.

We have established both directions so we have completed the proof.

Read the proof above very carefully. Notice that both directions had two parts: the existence of paths, and the uniqueness of paths (which related to the fact that there were no cycles). In this case, these two parts were really separate. In fact, if we just considered graphs with no cycles (a forest), then we could still do the parts of the proof that explore the uniqueness of paths between vertices, even if there might not exist paths between vertices.

This observation allows us to state the following *corollary:* [5]

Corollary 5.2.2 *A graph F is a forest if and only if between any pair of vertices in F there is at most one path.*

We do not give a proof of the corollary (it is, after all, supposed to follow directly from the proposition) but for practice, you are asked to give a careful proof in the exercises. When you do so, try to use proof by contrapositive instead of proof by contradiction.

Our second proposition tells us that all trees have **leaves:** vertices of degree one.

Proposition 5.2.3 *Any tree with at least two vertices has at least two vertices of degree one.*

Proof. We give a proof by contradiction. Let *T* be a tree with at least two vertices, and suppose, contrary to stipulation, that there are not two vertices of degree one.

Let *P* be a path in *T* of longest possible length. Let u and v be the endpoints of the path. Since T does not have two vertices of degree one, at least one of these must have degree two or higher. Say that it is u. We know that *u* is adjacent to a vertex in the path P, but now it must also be adjacent to another vertex, call it *u'*.

Where is u'? It cannot be a vertex of P, because if it was, there would be two distinct paths from u to u': the edge between them, and the first part of P (up to u'). But u' also cannot be outside of P, for if it was, there would be a path from u' to v that was longer than P, which has longest possible length.

This contradiction proves that there must be at least two vertices of degree one. In fact, we can say a little more: u and v must *both* have degree one.

The proposition is quite useful when proving statements about trees, because we often prove statements about trees by induction. To do so, we need to reduce a given tree to a smaller tree (so we can apply the inductive hypothesis). Getting rid of a vertex of degree one is an obvious choice, and now we know there is always one to get rid of.

To illustrate how induction is used on trees, we will consider the relationship between the number of vertices and number of edges in trees. Is there a tree with exactly 7 vertices and 7 edges? Try to draw one. Could a tree with 7 vertices have only 5 edges? There is a good reason that these seem impossible to draw.

Proposition 5.2.4 *Let T be a tree with v vertices and e edges. Then e = v - 1.*

Proof. We will give a proof by induction on the number of vertices in the tree. That is, we will prove that every tree with v vertices has exactly $v-1$ edges, and then use induction to show this is true for all $v \geq 1$.

For the base case, consider all trees with $v = 1$ vertices. There is only one such tree: the graph with a single isolated vertex. This graph has $e = 0$ edges, so we see that $e = v-1$ as needed. Now for the inductive case, fix $k \geq 1$ and assume that all trees with $v = k$ vertices have exactly $e = k-1$ edges. Now consider an arbitrary tree T with $v = k + 1$ vertices. By **Proposition 5.2.3,** T has a vertex v_0 of degree one. Let T' be the tree resulting from removing v_0 from T (together with its incident edge). Since we removed a leaf, T' is still a tree (the unique paths between pairs of vertices in TI are the same as the unique paths between them in T).

Now T' has k vertices, so by the inductive hypothesis, has $k-1$ edges. What can we say about T? Well, it has one more edge than T', so it has k edges. But this is exactly what we wanted: $v = k + 1$, $e = k$ so indeed $e = v - 1$. This completes the inductive case, and the proof.

There is a very important feature of this induction proof that is worth noting. Induction makes sense for proofs about graphs because we can think of graphs as growing into larger graphs. However, this does NOT work. It would not be correct to start with a tree with k vertices, and then add a new vertex and edge to get a tree with $k + 1$ vertices, and note that the number of edges also grew by one. Why is this bad? Because how do you know that every tree with $k + 1$ vertices is the result of adding a vertex to your arbitrary starting tree? You don't!

The point is that whenever you give an induction proof that a statement about graphs that holds for all graphs with v vertices, you must start with an arbitrary graph with v + 1 vertices, then reduce that graph to a graph with v vertices, to which you can apply your inductive hypothesis.

Rooted Trees

So far, we have thought of trees only as a particular kind of graph. However, it is often useful to add additional structure to trees to help solve problems. Data is often structured like a tree. This book, for example, has a tree structure: draw a vertex for the book itself. Then draw vertices for each chapter, connected to the book vertex. Under each chapter, draw a vertex for each section, connecting it to the chapter it belongs to. The graph will not have any cycles; it will be a tree. But a tree with clear hierarchy which is not present if we don't identify the book vertex as the "top".

As soon as one vertex of a tree is designated as the **root**, then every other vertex on the tree can be characterized by its position relative to the root. This works because there is a unique path between any two vertices in a tree. So from any vertex, we can travel back to the root in exactly one way. This also allows us to describe how distinct vertices in a rooted tree are related.

If two vertices are adjacent, then we say one of them is the **parent** of the other, which is called the child of the parent. Of the two, the parent is the vertex that is closer to the root. Thus the root of a tree is a parent, but is not the child of any vertex (and is unique in this respect: all non-root vertices have *exactly one* parent).

Not surprisingly, the child of a child of a vertex is called the **grandchild** of the vertex (and it is the **grandparent**). More in general, we say that a vertex v is a **descendent** of a vertex u provided u is a vertex on the path from v to the root. Then we would call u an ancestor of v.

For most trees (in fact, all except paths with one end the root), there will be pairs of vertices neither of which is a descendant of the other. We might call these cousins or siblings. In fact, vertices u and v are called siblings provided they have the same parent. Note that siblings are never adjacent (do you see why?).

Example 5.2.5

Consider the tree below.

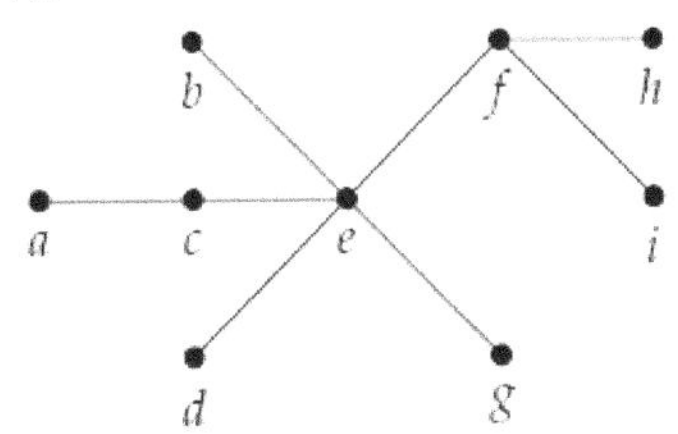

If we designate vertex *f* as the root, then *e, h,* and *i* are the children of *f*, and are siblings of each other. Among the other things we cay say are that a is a child of c, and a descendant of *f*. The vertex g is a descendant of *f*, in fact, is a grandchild of *f*. Vertices g and d are siblings, since they have the common parent *e*.

Notice how this changes if we pick a different vertex for the root. If a is the root, then its lone child is c, which also has only one child, namely e. We would then have *f* the child of e (instead of the other way around), and f is the descendant of a, instead of the ancestor. *f* and g are now siblings.

All of this flowery language helps us describe how to navigate through a tree. Traversing a tree, visiting each vertex in some order, is a key step in many algorithms. Even if the tree is not rooted, we can always form a rooted tree by picking any vertex as the root. Here is an example of why doing so can be helpful.

Example 5.2.6

Explain why every tree is a bipartite graph.

Solution

To show that a graph is bipartite, we must divide the vertices into two sets *A* and *B* so that no two vertices in the same set are adjacent. Here is an algorithm that does just this.

Designate any vertex as the root. Put this vertex in set *A*. Now put all of the children of the root in set *B*. None of these children are adjacent (they are siblings), so we are good so far. Now put into *A* every child of every vertex in *B* (i.e., every grandchild of the root). Keep going until all vertices have been assigned one of the sets, alternating between *A* and *B* every "generation." That is, a vertex is in set *B* if and only if it is the child of a vertex in set *A*.

The key to how we partitioned the tree in the example was to know which vertex to assign to a set next. We chose to visit all vertices in the same generation before any vertices of the next generation. This is usually called a **breadth first search** (we say "search" because you often traverse a tree looking for vertices with certain properties).

In contrast, we could also have partitioned the tree in a different order. Start with the root, put it in A. Then look for one child of the root to put in *B*. Then find a child of that vertex, into A, and then find its child, into B, and so on. When you get to a vertex with no children, retreat to its parent and see if the parent has any other children. So we travel as far from the root as fast as possible, then backtrack until we can move forward again. This is called **depth first search.**

These algorithmic explanations can serve as a proof that every tree is bipartite, although care needs to be spent to prove that the algorithms are correct. Another approach to prove that all trees are bipartite, using induction, is requested in the exercises.

Spanning Trees

One of the advantages of trees is that they give us a few simple ways to travel through the vertices. If a connected graph is not a tree, then we can still use these traversal algorithms if we identify a subgraph that *is* a tree. First we should consider if this even makes sense. Given any connected graph G, will there always be a subgraph that is a tree? Well, that is actually too easy: you could just take a single vertex of *G*. If we want to use this subgraph to tell us how to visit all vertices, then we want our subgraph to include all of the vertices. We call such a tree a **spanning tree.**

It turns out that every connected graph has one (and usually many).

Spanning tree

Given a connected graph *G*, a **spanning tree** of *G* is a subgraph of *G* which is a tree and includes all the vertices of *G*.

Every connected graph has a spanning tree.

How do we know? We can give an algorithm for *finding* a spanning tree! Start with a connected graph G. If there is no cycle, then G is already a tree and we are done. If there is a cycle, let e be any edge in that cycle and consider the new graph $G_1 = G - e$ (i.e., the graph you get by deleting e). This tree is still connected since e belonged to a cycle, there were at least two paths between its incident vertices. Now repeat: if G_1 has no cycles, we are done, otherwise define G_2 to be $G_1 - e_1$, where e_1 is an edge in a cycle in G_1. Keep going. This process must eventually stop, since there are only a finite number of edges to remove. The result will be a tree, and since we never removed any vertex, a *spanning* tree.

This is by no means the only algorithm for finding a spanning tree. You could have started with the empty graph and added edges that belong to G as long as adding them would not create a cycle. You have some choices as to which edges you add first: you could always add an edge adjacent to edges you have already added (after the first one, of course), or add them using some other order. Which spanning tree you end up with depends on these choices.

Example 5.2.7

Find two different spanning trees of the graph,

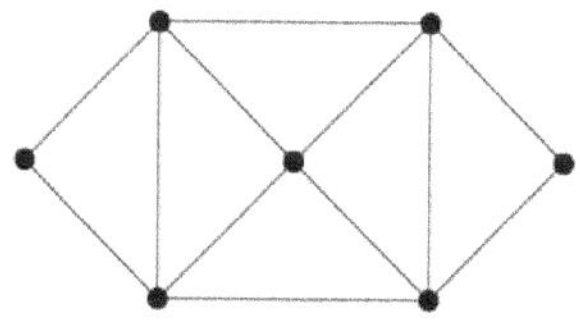

Solution. Here are two spanning trees.

Although we will not consider this in detail, these algorithms are usually applied to *weighted* graphs. Here every edge has some weight or cost assigned to it. The goal is to find a spanning tree that has the smallest possible combined weight. Such a tree is called a **minimum spanning tree**. Finding the minimum spanning tree uses basically the same algorithms as we described above, but when picking an edge to add, you always pick the smallest (or when removing an edge, you always remove the largest).[6]

Exercises

1. **Which of the following graphs are trees?**

   (a) $G = (V, E)$ with $V = \{a, b, c, d, e\}$ and $E = \{\{a, b\}, \{a, e\}, \{b, c\}, \{c, d\}, \{d, e\}\}$

   (b) $G = (V, E)$ with $V = \{a, b, c, d, e\}$ and $E = \{\{a, b\}, \{b, c\}, \{c, d\}, \{d, e\}\}$

   (c) $G = (V, E)$ with $V = \{a, b, c, d, e\}$ and $E = \{\{a, b\}, \{a, c\}, \{a, d\}, \{a, e\}\}$

   (d) $G = (V, E)$ with $V = \{a, b, c, d, e\}$ and $E = \{\{a, b\}, \{a, c\}, \{d, e\}\}$

2. **For each degree sequence below, decide whether it must always, must never, or could possibly be a degree sequence for a tree. Remember, a degree sequence lists out the degrees (number of edges incident to the vertex) of all the vertices in a graph in non-increasing order.**

   (a) (4, 1, 1, 1, 1)　(b) (3, 3, 2, 1, 1)

   (c) (2, 2, 2, 1, 1)　(d) (4, 4, 3, 3, 3, 2, 2, 1, 1, 1, 1, 1, 1, 1)

3. **For each degree sequence below, decide whether it must always, must never, or could possibly be a degree sequence for a tree. Justify your answers.**

   (a) (3, 3, 2, 2, 2)　(b) (3, 2, 2, 1, 1, 1)

   (c) (3, 3, 3, 1, 1, 1)　(d) (4, 4, 1, 1, 1, 1, 1, 1)

4. **Suppose you have a graph with v vertices and e edges that satisfies $v = e + 1$. Must the graph be a tree? Prove your answer.**

5. **Prove that any graph (not necessarily a tree) with v vertices and e edges that satisfies $v > e + 1$ will NOT be connected.**

6. **If a graph G with v vertices and e edges is connected and has $v < e + 1$, must it contain a cycle? Prove your answer.**

7. **We define a forest to be a graph with no cycles.**

   (a) Explain why this is a good name. That is, explain why a forest is a union of trees.

   (b) Suppose F is a forest consisting of m trees and v vertices. How many edges does F have? Explain.

   (c) Prove that any graph G with v vertices and e edges that satisfies $v < e + 1$ must contain a cycle (i.e., not be a forest).

8. **Give a careful proof of Corollary 4.2.2: A graph is a forest if and only if there is at most one path between any pair of vertices. Use proof by contrapositive (and not a proof by contradiction) for both directions.**

9. **Give a careful proof by induction on the number of vertices, that every tree is bipartite.**

10. **Consider the tree drawn below.**

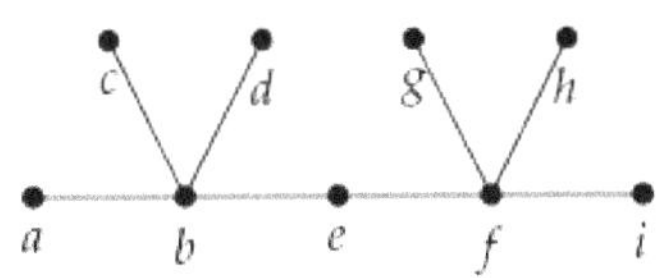

   (a) Suppose we designate vertex *e* as the root. List the children, parents and siblings of each vertex. Does any vertex other than *e* have grandchildren?

   (b) Suppose *e* is *not* chosen as the root. Does our choice of root vertex change the *number* of children e has? The number of grandchildren? How many are there of each?

   (c) In fact, pick any vertex in the tree and suppose it is not the root. Explain why the number of children of that vertex does not depend on which other vertex is the root.

   (d) Does the previous part work for other trees? Give an example of a different tree for which it holds. Then either prove that it always holds or give an example of a tree for which it doesn't.

11. **Let T be a rooted tree that contains vertices u, v, and w (among possibly others). Prove that if w is a descendant of both u and v, then u is a descendant of v or v is a descendant of u.**

12. **Unless it is already a tree, a given graph G will have multiple spanning trees. How similar or different must these be?**

(a) Must all spanning trees of a given graph be isomorphic to each other? Explain why or give a counterexample.

(b) Must all spanning trees of a given graph have the same number of edges? Explain why or give a counterexample.

(c) Must all spanning trees of a graph have the same number of leaves (vertices of degree 1)? Explain why or give a counterexample.

13. Find all spanning trees of the graph below. How many different spanning trees are there? How many different spanning trees are there *up to isomorphism* (that is, if you grouped all the spanning trees by which are isomorphic, how many groups would you have)?

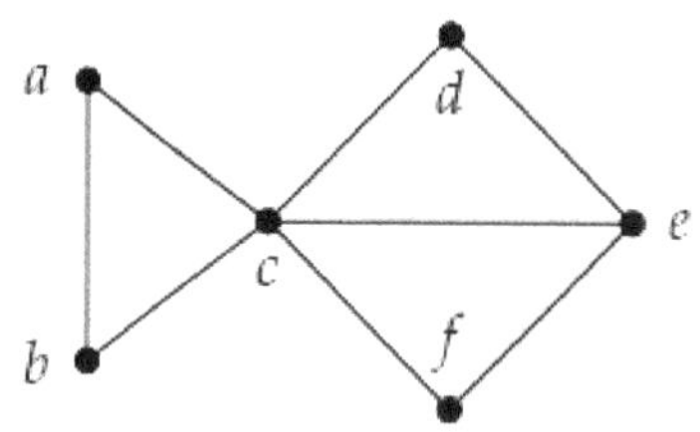

14. Give an example of a graph that has exactly 7 different spanning trees. Note, it is acceptable for some or all of these spanning trees to be isomorphic.

15. Prove that every connected graph which is not itself a tree must have at last three different (although possibly isomorphic) spanning trees.

16. Consider edges that must be in every spanning tree of a graph. Must every graph have such an edge? Give an example of a graph that has exactly one such edge.

5.3 PLANAR GRAPHS

Investigate!

When a connected graph can be drawn without any edges crossing, it is called **planar.** When a planar graph is drawn in this way, it divides the plane into regions called **faces.**

Draw, if possible, two different planar graphs with the same number of vertices, edges, and faces.

Draw, if possible, two different planar graphs with the same number of vertices and edges, but a different number of faces.

Attempt the above activity before proceeding

When is it possible to draw a graph so that none of the edges cross? If this is possible, we say the graph is **planar** (since you can draw it on the *plane*).

Notice that the definition of planar includes the phrase "it is possible to." This means that even if a graph does not look like it is planar, it still might be. Perhaps you can redraw it in a way in which no edges cross. For example, this is a planar graph:

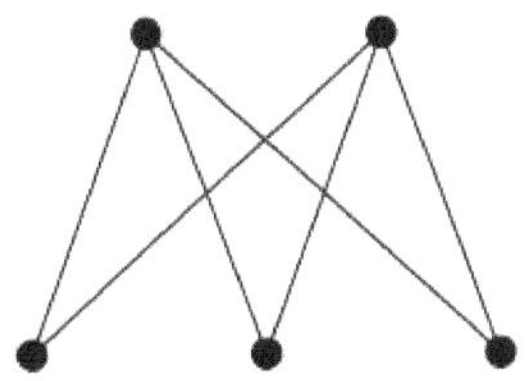

That is because we can redraw it like this:

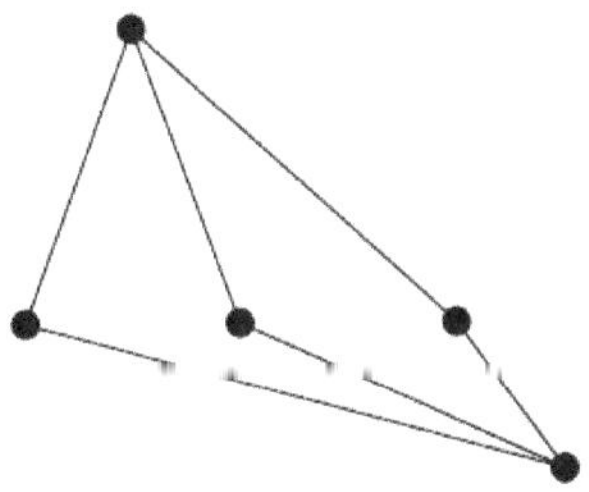

The graphs are the same, so if one is planar, the other must be too. How- ever, the original drawing of the graph was not a **planar representation** of the graph.

When a planar graph is drawn without edges crossing, the edges and vertices of the graph divide the plane into regions. We will call each region a **face.** The graph above has 3 faces (yes, we *do* include the "outside" region as a face). The number of faces does not change no matter how you draw the graph (as long as you do so without the edges crossing), so it makes sense to ascribe the number of faces as a property of the planar graph.

WARNING: you can only count faces when the graph is drawn in a planar way. For example, consider these two representations of the same graph:

If you try to count faces using the graph on the left, you might say there are 5 faces (including the outside). But drawing the graph with a planar representation shows that in fact there are only 4 faces.

There is a connection between the number of vertices (v), the number of edges (e) and the number of faces (f) in any connected planar graph. This relationship is called Euler's formula.

Euler's Formula for Planar Graphs.

For any connected planar graph with v vertices, e edges and f faces, we have

$$v - e + f = 2.$$

Why is Euler's formula true? One way to convince yourself of its validity is to draw a planar graph step by step. Start with the graph P_2:

Any connected graph (besides just a single isolated vertex) must contain this subgraph. Now build up to your graph by adding edges and vertices. Each step will consist of either adding a new vertex connected by a new edge to part of your graph (so creating a new "spike") or by connecting two vertices already in the graph with a new edge (completing a circuit).

What do these "moves" do? When adding the spike, the number of edges increases by 1, the number of vertices increases by one, and the number of faces remains the same. But this means that $v - e + f$ does not change. Completing a circuit adds one edge, adds one face, and keeps the number of vertices the same. So again, $v - e + f$ does not change.

Since we can build any graph using a combination of these two moves, and doing so never changes the quantity $v - e + f$, that quantity will be the same for all graphs. But notice that our starting graph P_2 has $v = 2$, $e = 1$ and $f = 1$, so $v - e + f = 2$. This argument is essentially a proof by induction. A good exercise would be to rewrite it as a formal induction proof.

Non-planar Graphs

Investigate!

For the complete graphs K_n, we would like to be able to say something about the number of vertices, edges, and (if the graph is planar) faces. Let's first consider K_3:

- How many vertices does K_3 have? How many edges?
- If K_3 is planar, how many faces should it have?
- Repeat parts (1) and (2) for K_4, K_5, and K_{23}.

What about complete bipartite graphs? How many vertices, edges, and faces (if it were planar) does $K_{7,4}$ have? For which values of m and n are K_n and $K_{m,n}$ planar?

Attempt the above activity before proceeding

Not all graphs are planar. If there are too many edges and too few vertices, then some of the edges will need to intersect. The smallest graph where this happens is K_5.

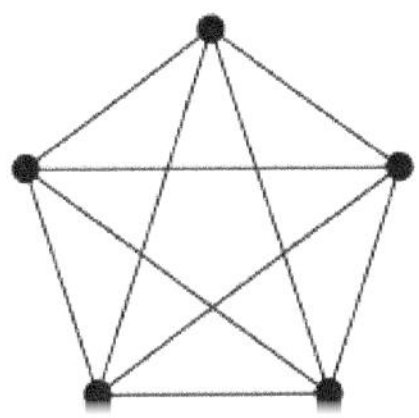

If you try to redraw this without edges crossing, you quickly get into trouble. There seems to be one edge too many. In fact, we can prove that no matter how you draw it, K_5 will always have edges crossing.

Theorem 5.3.1 K_5 is not planar.

Proof. The proof is by contradiction. So assume that K5 is planar. Then the graph must satisfy Euler's formula for planar graphs. K5 has 5 vertices and 10 edges, so we get

$$5 - 10 + f = 2,$$

which says that if the graph is drawn without any edges crossing, there would be $f = 7$ faces.

Now consider how many edges surround each face. Each face must be surrounded by at least 3 edges. Let B be the total number of boundaries around all the faces in the graph. Thus we have that $3f \leq B$. But also $B = 2e$, since each edge is used as a boundary exactly twice. Putting this together we get

$$3f \leq 2e.$$

But this is impossible, since we have already determined that $f = 7$ and $e = 10$, and $21 \nleq 20$. This is a contradiction so in fact K_5 is not planar.

The other simplest graph which is not planar is $K_{3,3}$

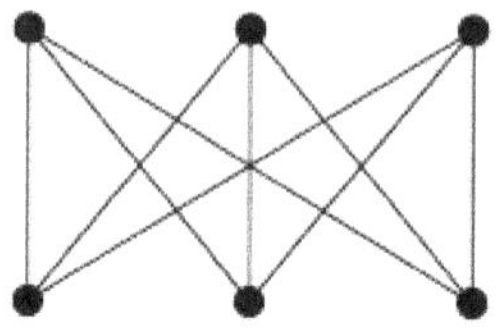

Proving that $K_{3,3}$ is not planar answers the houses and utilities puzzle: it is not possible to connect each of three houses to each of three utilities without the lines crossing.

Theorem 5.3.2 $K_{3,3}$ is not planar.

Proof. Again, we proceed by contradiction. Suppose $K_{3,3}$ were planar. Then by Euler's formula there will be 5 faces, since v = 6, e = 9, and 6− 9 + f = 2.

How many boundaries surround these 5 faces? Let B be this number. Since each edge is used as a boundary twice, we have B = 2e. Also, B ≥ 4 f since each face is surrounded by 4 or more boundaries. We know this is true because $K_{3,3}$ is bipartite, so does not contain any 3-edge cycles. Thus

$$4f \le 2e.$$

But this would say that 20 ≤ 18, which is clearly false. Thus $K_{3,3}$ is not planar.

Note the similarities and differences in these proofs. Both are proofs by contradiction, and both start with using Euler's formula to derive the (supposed) number of faces in the graph. Then we find a relationship between the number of faces and the number of edges based on how many edges surround each face. This is the only difference. In the proof for K_5, we got 3 f ≤ 2e and for $K_{3,3}$ we go 4 f ≤ 2e. The coefficient of f is the key. It is the smallest number of edges which could surround any face. If some number of edges surround a face, then these edges form a cycle. So that number is the size of the smallest cycle in the graph.

In general, if we let g be the size of the smallest cycle in a graph (g stands for girth, which is the technical term for this) then for any planar graph we have $gf \le 2e$. When this disagrees with Euler's formula, we know for sure that the graph cannot be planar.

Polyhedra

Investigate!

A cube is an example of a convex polyhedron. It contains 6 identical squares for its faces, 8 vertices, and 12 edges. The cube is a regular polyhedron (also known as a **Platonic solid**) because each face is an identical **regular polygon** and each vertex joins an equal number of faces.

There are exactly four other regular polyhedra: the tetrahedron, octahedron, dodecahedron, and icosahedron with 4, 8, 12 and 20 faces respectively. How many vertices and edges do each of these have?

Attempt the above activity before proceeding

Another area of mathematics where you might have heard the terms "vertex," "edge," and "face" is geometry. A **polyhedron** is a geometric solid made up of flat polygonal faces joined at edges and vertices. We are especially interested in **convex** polyhedra, which means that any line segment connecting two points on the interior of the polyhedron must be entirely contained inside the polyhedron.[7]

Notice that since 8− 12 + 6 = 2, the vertices, edges and faces of a cube satisfy Euler's formula for planar graphs. This is not a coincidence. We can represent a cube as a planar graph by projecting the vertices and edges onto the plane. One such projection looks like this:

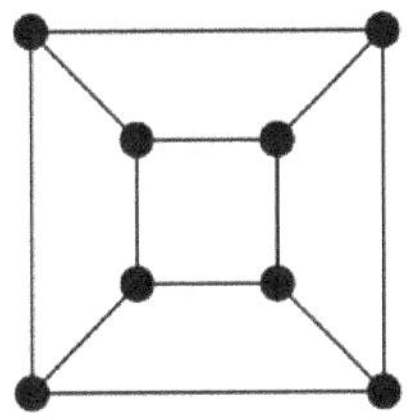

In fact, every convex polyhedron can be projected onto the plane without edges crossing. Think of placing the polyhedron inside a sphere, with a light at the center of the sphere. The edges and vertices of the polyhedron cast a shadow onto the interior of the sphere. You can then cut a hole in the sphere in the middle of one of the projected faces and "stretch" the sphere to lie down flat on the plane. The face that was punctured becomes the "outside" face of the planar graph.

The point is, we can apply what we know about graphs (in particular planar graphs) to convex polyhedra. Since every convex polyhedron can be represented as a planar graph, we see that Euler's formula for planar graphs holds for all convex polyhedra as well. We also can apply the same sort of reasoning we use for graphs in other contexts to convex polyhedra. For example, we know that there is no convex polyhedron with 11 vertices all of degree 3, as this would make 33/2 edges.

Example 5.3.3

Is there a convex polyhedron consisting of three triangles and six pentagons? What about three triangles, six pentagons and five heptagons (7-sided polygons)?

Solution

How many edges would such polyhedra have? For the first proposed polyhedron, the triangles would contribute a total of 9 edges, and the pentagons would contribute 30. However, this counts each edge twice (as each edge borders exactly two faces), giving 39/2 edges, an impossibility. There is no such polyhedron.

The second polyhedron does not have this obstacle. The extra 35 edges contributed by the heptagons give a total of 74/2 = 37 edges. So far so good. Now how many vertices does this supposed polyhedron have? We can use Euler's formula. There are 14 faces, so we have $v - 37 + 14 = 2$ or equivalently $v = 25$. But now use the vertices to count the edges again. Each vertex must have degree at *least* three (that is, each vertex joins at least three faces since the interior angle of all the polygons must be less that 180°), so the sum of the degrees of vertices is at least 75. Since the sum of the degrees must be exactly twice the number of edges, this says that there are strictly more than 37 edges. Again, there is no such polyhedron.

To conclude this application of planar graphs, consider the regular polyhedra. We claimed there are only five. How do we know this is true? We can prove it using graph theory.

Theorem 5.3.4 There are exactly five regular polyhedra.

Proof. Recall that all the faces of a regular polyhedron are identical regular polygons, and that each vertex has the same degree. Consider four cases, depending on the type of regular polygon.

Case 1: Each face is a triangle. Let f be the number of faces. There are then $3f/2$ edges. Using Euler's formula we have $v - 3f/2 + f = 2$ so $v = 2 + f/2$. Now each vertex has the same degree, say k. So the number of edges is also $kv/2$. Putting this together gives

$$e = \frac{3f}{2} = \frac{k(2 + f/2)}{2},$$

which says

$$k = \frac{6f}{4 + f}.$$

Both k and f must be positive integers. Note that $\frac{6f}{4+f}$ is an increasing function for positive f, bounded above by a horizontal asymptote at $k = 6$. Thus the only possible values for k are 3, 4, and 5. Each of these are possible. To get $k = 3$, we need $f = 4$ (this is the tetrahedron). For $k = 4$ we take $f = 8$ (the octahedron). For $k = 5$ take $f = 20$ (the icosahedron).

Thus there are exactly three regular polyhedra with triangles for faces.

Case 2: Each face is a square. Now we have $e = 4f/2 = 2f$. Using Euler's formula we get $v = 2 + f$, and counting edges using the degree k of each vertex gives us

$$e = 2f = \frac{k(2+f)}{2}.$$

Solving for k gives

$$k = \frac{4f}{2+f} = \frac{8f}{4+2f}.$$

This is again an increasing function, but this time the horizontal asymptote is at $k = 4$, so the only possible value that k could take is 3. This produces 6 faces, and we have a cube. There is only one regular polyhedron with square faces.

Case 3: Each face is a pentagon. We perform the same calculation as above, this time getting $e = 5f/2$ so $v = 2 + 3f/2$. Then

$$e = \frac{5f}{2} = \frac{k(2+3f/2)}{2},$$

so

$$k = \frac{10f}{4+3f}.$$

Now the horizontal asymptote is at $\frac{10}{3}$. This is less than 4, so we can only hope of making $k = 3$. We can do so by using 12 pentagons, getting the dodecahedron. This is the only regular polyhedron with pentagons as faces.

Case 4: Each face is an n-gon with $n \geq 6$. Following the same procedure as above, we deduce that

$$k = \frac{2nf}{4+(n-2)f},$$

which will be increasing to a horizontal asymptote of $\frac{2n}{n-2}$. When n = 6, this asymptote is at $k = 3$. Any larger value of n will give an even smaller asymptote. Therefore no regular polyhedra exist with faces larger than pentagons.[8]

Exercises

1. **Is it possible for a planar graph to have 6 vertices, 10 edges and 5 faces? Explain.**
2. **The graph G has 6 vertices with degrees 2, 2, 3, 4, 4, 5. How many edges does G have? Could G be planar? If so, how many faces would it have. If not, explain.**
3. **Is it possible for a connected graph with 7 vertices and 10 edges to be drawn so that no edges cross and create 4 faces? Explain.**
4. **Is it possible for a graph with 10 vertices and edges to be a connected planar graph? Explain.**

5. **Is there a connected planar graph with an odd number of faces where every vertex has degree 6? Prove your answer.**
6. **I'm thinking of a polyhedron containing 12 faces. Seven are triangles and four are quadralaterals. The polyhedron has 11 vertices including those around the mystery face. How many sides does the last face have?**
7. **Consider some classic polyhedrons.**
   (a) An *octahedron* is a regular polyhedron made up of 8 equilateral triangles (it sort of looks like two pyramids with their bases glued together). Draw a planar graph representation of an octahedron. How many vertices, edges and faces does an octahedron (and your graph) have?
   (b) The traditional design of a soccer ball is in fact a (spherical projection of a) truncated icosahedron. This consists of 12 regular pentagons and 20 regular hexagons. No two pentagons are adjacent (so the edges of each pentagon are shared only by hexagons). How many vertices, edges, and faces does a truncated icosahedron have? Explain how you arrived at your answers. Bonus: draw the planar graph representation of the truncated icosahedron.
   (c) Your "friend" claims that he has constructed a convex polyhedron out of 2 triangles, 2 squares, 6 pentagons and 5 octagons. Prove that your friend is lying. **Hint:** each vertex of a convex polyhedron must border at least three faces.
8. **Prove Euler's formula using induction on the number of edges in the graph.**
9. **Prove Euler's formula using induction on the number of vertices in the graph.**
10. **Euler's formula ($v - e + f = 2$) holds for all connected planar graphs. What if a graph is not connected? Suppose a planar graph has two components. What is the value of $v - e + f$ now? What if it has k components?**
11. **Prove that the Petersen graph (below) is not planar.**

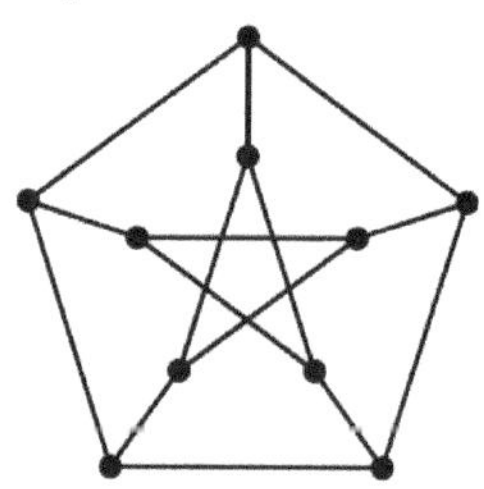

12. **Prove that any planar graph with v vertices and e edges satisfies $e \leq 3v - 6$.**
13. **Prove that any planar graph must have a vertex of degree 5 or less.**
14. **Give a careful proof that the graph below is not planar.**

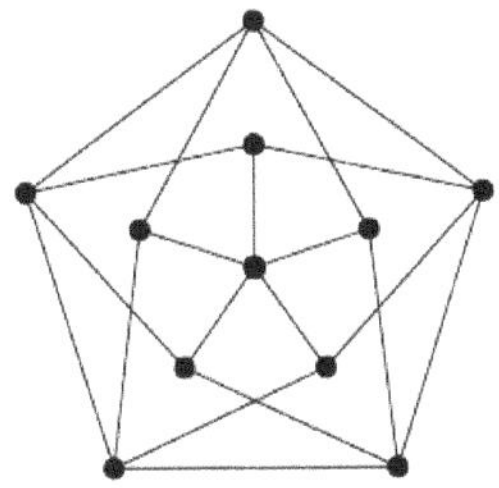

15. **Explain why we cannot use the same sort of proof we did in Exercise 4.3.14 to prove that the graph below is not planar. Then explain how you know the graph is not planar anyway.**

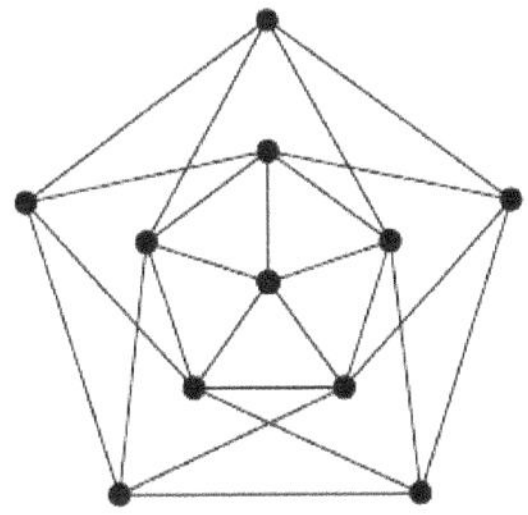

5.4 COLORING

Investigate!

Mapmakers in the fictional land of Euleria have drawn the borders of the various dukedoms of the land. To make the map pretty, they wish to color each region. Adjacent regions must be colored differently, but it is perfectly fine to color two distant regions with the same color. What is the fewest colors the mapmakers can use and still accomplish this task?

Attempt the above activity before proceeding

- Perhaps the most famous graph theory problem is how to color maps.
- Given any map of countries, states, counties, etc., how many colors are needed to color each region on the map so that neighboring regions are colored differently?

- Actual map makers usually use around seven colors. For one thing, they require watery regions to be a specific color, and with a lot of colors it is easier to find a permissible coloring. We want to know whether there is a smaller palette that will work for any map.
- How is this related to graph theory? Well, if we place a vertex in the center of each region (say in the capital of each state) and then connect two vertices if their states share a border, we get a graph. Coloring regions on the map corresponds to coloring the vertices of the graph. Since neighboring regions cannot be colored the same, our graph cannot have vertices colored the same when those vertices are adjacent.
- In general, given any graph G, a coloring of the vertices is called (not surprisingly) a **vertex coloring.** If the vertex coloring has the property that adjacent vertices are colored differently, then the coloring is called **proper.** Every graph has a proper vertex coloring. For example, you could color every vertex with a different color. But often you can do better. The smallest number of colors needed to get a proper vertex coloring is called the **chromatic number** of the graph, written $\chi(G)$.

Example 5.4.1

Find the chromatic number of the graphs below.

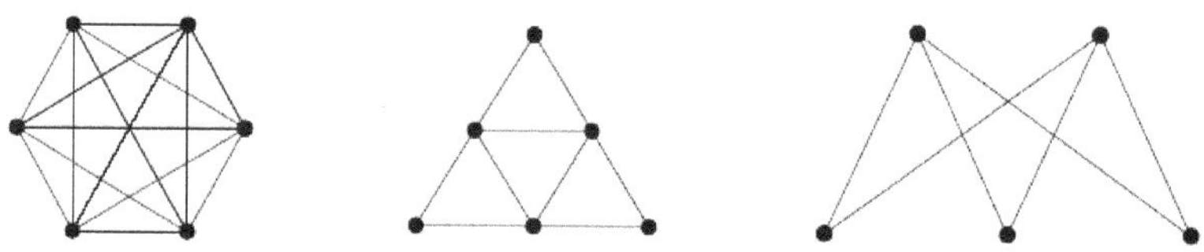

Solution

The graph on the left is K_6. The only way to properly color the graph is to give every vertex a different color (since every vertex is adjacent to every other vertex). Thus the chromatic number is 6.

The middle graph can be properly colored with just 3 colors (Red, Blue, and Green). For example:

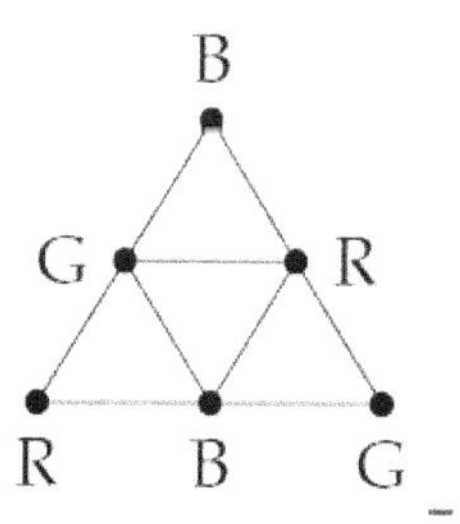

There is no way to color it with just two colors, since there are three vertices mutually adjacent (i.e., a triangle). Thus the chromatic number is 3.

The graph on the right is just $K_{2,3}$. As with all bipartite graphs, this graph has chromatic number 2: color the vertices on the top row red and the vertices on the bottom row blue.

It appears that there is no limit to how large chromatic numbers can get. It should not come as a surprise that Kn has chromatic number n. So how could there possibly be an answer to the original map coloring question? If the chromatic number of graph can be arbitrarily large, then it seems like there would be no upper bound to the number of colors needed for any map. But there is.

The key observation is that while it is true that for any number n, there is a graph with chromatic number n, only some graphs arrive as representations of maps. If you convert a map to a graph, the edges between vertices correspond to borders between the countries. So you should be able to connect vertices in such a way where the edges do not cross. In other words, the graphs representing maps are all *planar*!

So the question is, what is the largest chromatic number of any planar graph? The answer is the best known theorem of graph theory:

Theorem 5.4.2 The Four Color Theorem. *If G is a planar graph, then the chromatic number of G is less than or equal to 4. Thus any map can be properly colored with 4 or fewer colors.*

We will not prove this theorem. Really. Even though the theorem is easy to state and understand, the proof is not. In fact, there is currently no "easy" known proof of the theorem. The current best proof still requires powerful computers to check an *unavoidable set* of 633 *reducible configurations.* The idea is that every graph must contain one of these reducible configurations (this fact also needs to be checked by a computer) and that reducible configurations can, in fact, be colored in 4 or fewer colors.

Coloring in General

Investigate!

The math department plans to offer 10 classes next semester. Some classes cannot run at the same time (perhaps they are taught by the same professor, or are required for seniors).

Class:	Conflicts with:
A	D I
B	D I J
C	E F I
D	A B F
E	C H I
F	C D I
G	J
H	E I J
I	A B C E F H
J	B G H

How many different time slots are needed to teach these classes (and which should be taught at the same time)? More importantly, how could we use graph coloring to answer this question?

Attempt the Above Activity Before Proceeding

Cartography is certainly not the only application of graph coloring. There are plenty of situations in which you might wish to partition the objects in question so that related objects are not in the same set. For example, you might wish to store chemicals safely. To avoid explosions, certain pairs of chemicals should not be stored in the same room. By color- ing a graph (with vertices representing chemicals and edges representing potential negative interactions), you can determine the smallest number of rooms needed to store the chemicals.

Here is a further example:

Example 5.4.3

Radio stations broadcast their signal at certain frequencies. How- ever, there are a limited number of frequencies to choose from, so nationwide many stations use the same frequency. This works because the stations are far enough apart that their signals will not interfere; no one radio could pick them up at the same time.

Suppose 10 new radio stations are to be set up in a currently unpopulated (by radio stations) region. The radio stations that are close enough to each other to cause interference are recorded in the table below. What is the fewest number of frequencies the stations could use.

	KQEA	KQEB	KQEC	KQED	KQEE	KQEF	KQEG	KQEH	KQEI	KQEJ
KQEA			x			x	x			x
KQEB			x	x						
KQEC	x					x	x			x

KQED		x			x	x		x		
KQEE				x					x	
KQEF	x		x	x			x			x
KQEG	x		x			x				x
KQEH				x					x	
KQEI					x			x		x
KQEJ	x		x			x	x		x	

Solution

Represent the problem as a graph with vertices as the stations and edges when two stations are close enough to cause interference. We are looking for the chromatic number of the graph. Vertices that are colored identically represent stations that can have the same frequency.

This graph has chromatic number 5. A proper 5-coloring is shown on the right. Notice that the graph contains a copy of the complete graph K_5 so no fewer than 5 colors can be used.

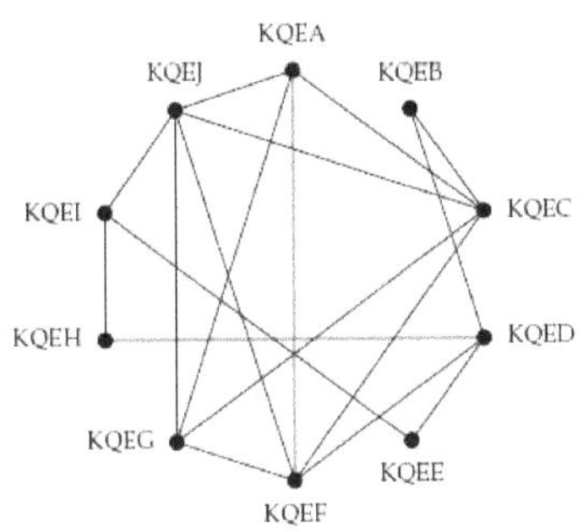

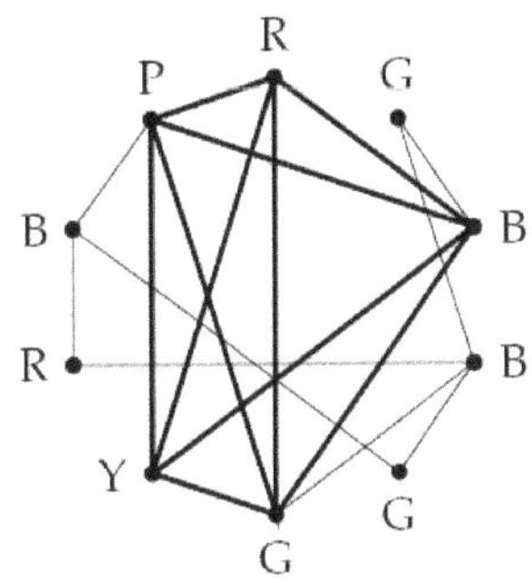

In the example above, the chromatic number was 5, but this is not a counterexample to the **Four Color Theorem 5.4.2,** since the graph representing the radio stations is not planar. It would be nice to have some quick way to find the chromatic number of a (possibly non-planar) graph. It turns out nobody knows whether an efficient algorithm for computing chromatic numbers exists.

While we might not be able to find the exact chromatic number of graph easily, we can often give a reasonable range for the chromatic number. In other words, we can give upper and lower bounds for chromatic number.

This is actually not very difficult: for every graph G, the chromatic number of G is at least 1 and at most the number of vertices of G.

What? You want better bounds on the chromatic number? Well you are in luck.

A **clique** in a graph is a set of vertices all of which are pairwise adjacent. In other words, a clique of size *n* is just a copy of the complete graph K_n. We define the **clique number** of a graph to be the largest *n* for which the graph contains a clique of size n. Any clique of size n cannot be colored with fewer than *n* colors, so we have a nice lower bound:

Theorem 5.4.4 *The chromatic number of a graph G is at least the clique number of G.*

There are times when the chromatic number of G is equal to the clique number. These graphs have a special name; they are called perfect. If you know that a graph is perfect, then finding the chromatic number is simply a matter of searching for the largest clique.9 However, not all graphs are perfect.

For an upper bound, we can improve on "the number of vertices" by looking to the degrees of vertices. Let Δ (G) be the largest degree of any vertex in the graph G. One reasonable guess for an upper bound on the chromatic number is χ (G) ≤ Δ (G) + 1. Why is this reasonable? Starting with any vertex, it together with all of its neighbors can always be colored in Δ(G)+1 colors, since at most we are talking about Δ(G)+1 vertices in this set. Now fan out! At any point, if you consider an already colored vertex, some of its neighbors might be colored, some might not. But no matter what, that vertex and its neighbors could all be colored distinctly, since there are at most Δ (G) neighbors, plus the one vertex being considered.

In fact, there are examples of graphs for which χ (G) = Δ (G) + 1. For any n, the complete graph K_n has chromatic number n, but Δ (K_n) = n– 1 (since every vertex is adjacent to every other vertex). Additionally, any *odd* cycle will have chromatic number 3, but the degree of every vertex in a cycle is 2. It turns out that these are the only two types of examples where we get equality, a result known as Brooks' Theorem.

Theorem 5.4.5 Brooks' Theorem. *Any graph G satisfies $\chi(G) \leq \Delta(G)$, unless G is a complete graph or an odd cycle, in which case $\chi(G) = \Delta(G) + 1$.*

The proof of this theorem is just complicated enough that we will not present it here (although you are asked to prove a special case in the exercises). The adventurous reader is encouraged to find a book on graph theory for suggestions on how to prove the theorem.

Coloring Edges

The chromatic number of a graph tells us about coloring vertices, but we could also ask about coloring edges. Just like with vertex coloring, we might insist that edges that are adjacent must be colored differently. Here, we are thinking of two edges as being adjacent if they are incident to the same vertex. The least number of colors required to properly color the edges of a graph G is called the **chromatic index** of G, written $\chi'(G)$.

Example 5.4.6

Six friends decide to spend the afternoon playing chess. Everyone will play everyone else once. They have plenty of chess sets but nobody wants to play more than one game at a time. Games will last an hour (thanks to their handy chess clocks). How many hours will the tournament last?

Solution

Represent each player with a vertex and put an edge between two players if they will play each other. In this case, we get the graph K_6:

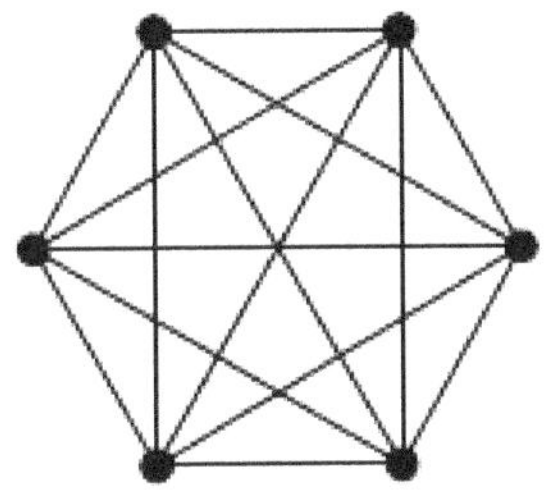

We must color the edges; each color represents a different hour. Since different edges incident to the same vertex will be colored differently, no player will be playing two different games (edges) at the same time. Thus we need to know the chromatic index of K_6.

- Notice that for sure $\chi'(K_6) \geq 5$, since there is a vertex of degree 5.
- It turns out 5 colors is enough (go find such a coloring). Therefore the friends will play for 5 hours.
- Interestingly, if one of the friends in the above example left, the remaining 5 chess-letes would still need 5 hours: the chromatic index of K_5 is also 5.
- In general, what can we say about chromatic index? Certainly $\chi'(G) \geq \Delta(G)$. But how much higher could it be? Only a little higher.

Theorem 5.4.7 Vizing's Theorem. *For any graph G, the chromatic index* $\chi I(G)$ is either $\Delta(G)$ or $\Delta(G) + 1$.

At first this theorem makes it seem like chromatic index might not be very interesting. However, deciding which case a graph is in is not always easy. Graphs for which $\chi'(G) = \Delta(G)$ are called class 1, while the others are called class 2. Bipartite graphs always satisfy $\chi'(G) = \Delta(G)$, so are class 1 (this was proved by König in 1916, decades before Vizing proved his theorem in 1964). In 1965 Vizing proved that all planar graphs with $\Delta(G) \geq 8$ are of class 1, but this does not hold for all planar graphs with $2 \leq \Delta(G) \leq 5$. Vizing conjectured that all planar graphs with $\Delta(G) = 6$ or

$\Delta(G) = 7$ are class 1; the $\Delta(G) = 7$ case was proved in 2001 by Sanders and Zhao; the $\Delta(G) = 6$ case is still open.

Ramsey Theory

There is another interesting way we might consider coloring edges, quite different from what we have discussed so far. What if we colored every edge of a graph either red or blue. Can we do so without, say, creating a *monochromatic* triangle (i.e., an all red or all blue triangle)? Certainly for some graphs the answer is yes. Try doing so for K_4. What about K_5? K_6? How far can we go?

The problem above is not too difficult and is a fun exercise. We could extend the question in a variety of ways. What if we had three colors? What if we were trying to avoid other graphs. Surprisingly, very little is known about these questions. For example, we know that you need to go up to K_{17} in order to force a monochromatic triangle using three colors, but nobody knows how big you need to go with more colors. Similarly, we know that using two colors K_{18} is the smallest graph that forces a monochromatic copy of K_4, but the best we have to force a monochromatic K_5 is a range, somewhere from K_{43} to K_{49}. If you are interested in these sorts of questions, this area of graph theory is called Ramsey theory. Check it out.

EXERCISES

1. **What is the smallest number of colors you need to properly color the vertices of $K_{4,5}$? That is, find the chromatic number of the graph.**
2. **Draw a graph with chromatic number 6 (i.e., which requires 6 colors to properly color the vertices). Could your graph be planar? Explain.**
3. **Find the chromatic number of each of the following graphs.**

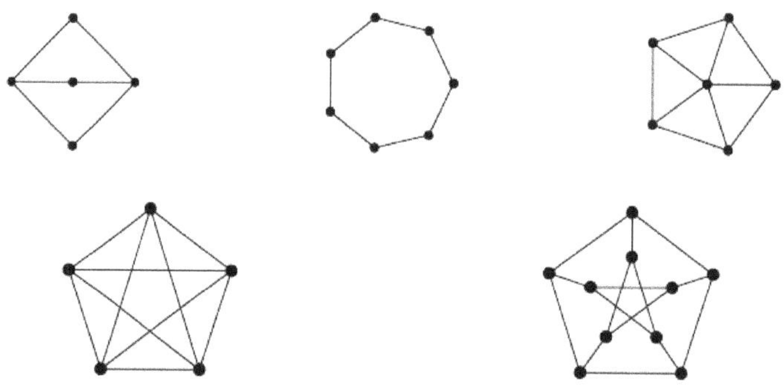

4. **A group of 10 friends decides to head up to a cabin in the woods (where nothing could possibly go wrong). Unfortunately, a number of these friends have dated each other in the past, and things are still a little awkward. To get to the cabin, they need to divide up into some number of cars, and no two people who dated should be in the same car.**

(a) What is the smallest number of cars you need if all the relation- ships were strictly heterosexual? Represent an example of such a situation with a graph. What kind of graph do you get?

(b) Because a number of these friends dated there are also conflicts between friends of the same gender, listed below. Now what is the smallest number of conflict-free cars they could take to the cabin?

Friend	A	B	C	D	E	F	G	H	I	J
Conflicts	CFJ	J	AEF	H	CFG	ACEGI	EFI	D	AFG	B

5. **What is the smallest number of colors that can be used to color the vertices of a cube so that no two adjacent vertices are colored identically?**

6. **Prove the chromatic number of any tree is two. Recall, a tree is a connected graph with no cycles.**

(a) Describe a procedure to color the tree below.

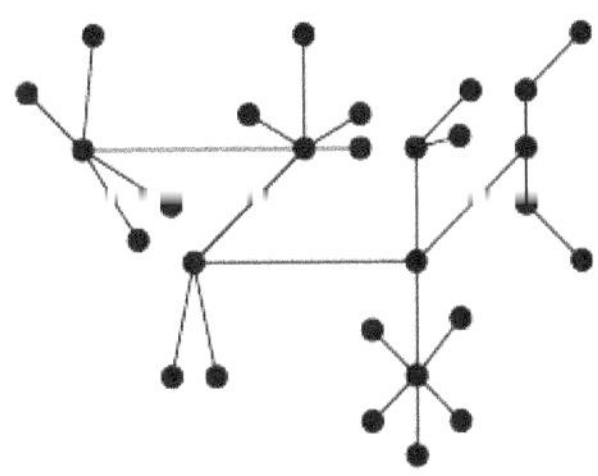

(b) The chromatic number of Cn is two when n is even. What goes wrong when n is odd?

(c) Prove that your procedure from part (a) always works for any tree.

(d) Now, prove using induction that every tree has chromatic number 2.

7. **The two problems below can be solved using graph coloring. For each problem, represent the situation with a graph, say whether you should be coloring vertices or edges and why, and use the coloring to solve the problem.**

(a) Your Quidditch league has 5 teams. You will play a tournament next week in which every team will play every other team once. Each team can play at most one match each day, but there is plenty of time in the day for multiple matches. What is the fewest number of days over which the tournament can take place?

(b) Ten members of Math Club are driving to a math conference in a neighboring state. However, some of these students have dated in the past, and things are still a little awkward. Each student lists which other students they refuse to share a car with; these conflicts are recorded in the table below. What is the fewest number of cars the club needs to make the trip? Do not worry about running out of seats, just avoid the conflicts.

Student	A	B	C	D	E	F	G	H	I	J
Conflicts	BEJ	ADG	HJ	BF	AI	DJ	B	CI	EHJ	ACFI

8. **Prove the 6-color theorem: every planar graph has chromatic number 6 or less. Do not assume the 4-color theorem (whose proof is MUCH harder), but you may assume the fact that every planar graph contains a vertex of degree at most 5.**

9. **Not all graphs are perfect. Give an example of a graph with chromatic number 4 that does not contain a copy of K_4. That is, there should be no 4 vertices all pairwise adjacent.**

10. **Find the chromatic number of the graph below and prove you are correct.**

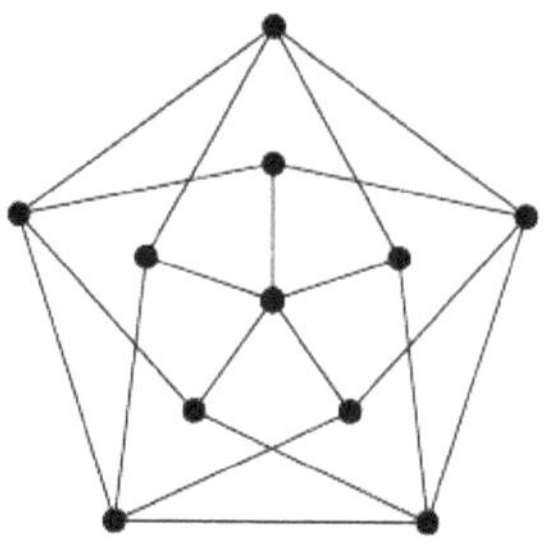

11. **Prove by induction on vertices that any graph G which contains at least one vertex of degree less than $\Delta(G)$ (the maximal degree of all vertices in G) has chromatic number at most $\Delta(G)$.**

12. **You have a set of magnetic alphabet letters (one of each of the 26 letters in the alphabet) that you need to put into boxes. For obvious reasons, you don't want to put two consecutive letters in the same box. What is the fewest number of boxes you need (assuming the boxes are able to hold as many letters as they need to)?**

13. **Suppose you colored edges of a graph either red or blue (not requiring that adjacent edges be colored differently). What must be true of the graph to guarantee some vertex is incident to three edges of the same color? Prove your answer.**

14. **Prove that if you color every edge of K6 either red or blue, you are guaranteed a monochromatic triangle (that is, an all red or an all blue triangle).**

5.5 EULER PATHS AND CIRCUITS

Investigate!

An **Euler path**, in a graph or multigraph, is a walk through the graph which uses every edge exactly once. An Euler circuit is an Euler path which starts and stops at the same vertex. Our goal is to find a quick way to check whether a graph (or multigraph) has an Euler path or circuit.

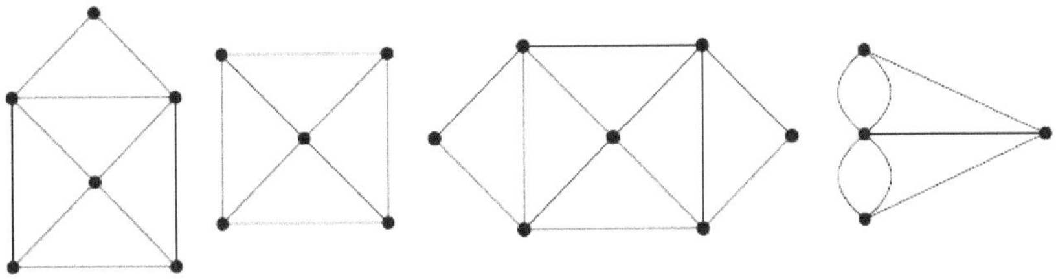

- Which of the graphs below have Euler paths? Which have Euler circuits?
- List the degrees of each vertex of the graphs above. Is there a connection between degrees and the existence of Euler paths and circuits?
- Is it possible for a graph with a degree 1 vertex to have an Euler circuit? If so, draw one. If not, explain why not. What about an Euler path?
- What if every vertex of the graph has degree 2. Is there an Euler path? An Euler circuit? Draw some graphs.
- Below is *part* of a graph. Even though you can only see some of the vertices, can you deduce whether the graph will have an Euler path or circuit?

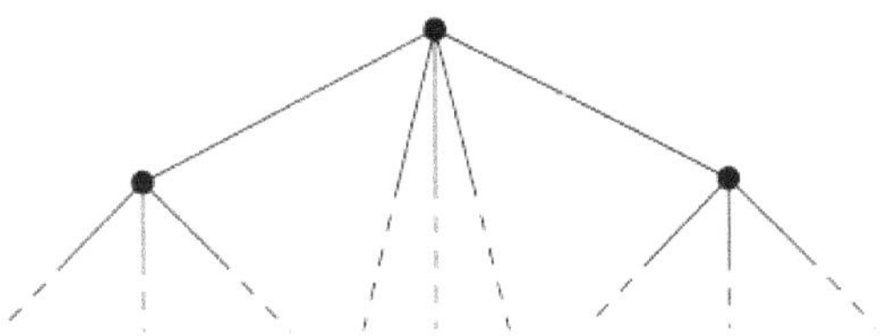

Attempt the above activity before proceeding

If we start at a vertex and trace along edges to get to other vertices, we create a *walk* through the graph. More precisely, a **walk** in a graph is a sequence of vertices such that every vertex in the sequence is adjacent to the vertices before and after it in the sequence. If the walk travels along every edge exactly once, then the walk is called an **Euler path** (or **Euler walk**). If, in addition, the starting and ending vertices are the same (so you trace along every edge exactly once and end up where you started), then the walk is called an **Euler circuit** (or **Euler tour**). Of course if a graph is not connected, there is no hope of finding such a path or circuit. For the rest of this section, assume all the graphs discussed are connected.

The bridges of Königsberg problem is really a question about the existence of Euler paths. There will be a route that crosses every bridge exactly once if and only if the graph below has an Euler path:

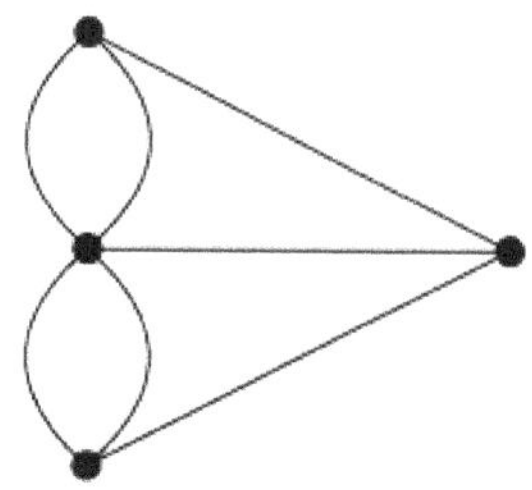

This graph is small enough that we could actually check every possible walk that does not reuse edges, and in doing so convince ourselves that there is no Euler path (let alone an Euler circuit). On small graphs which do have an Euler path, it is usually not difficult to find one. Our goal is to find a quick way to check whether a graph has an Euler path or circuit, even if the graph is quite large.

One way to guarantee that a graph does not have an Euler circuit is to include a "spike," a vertex of degree 1.

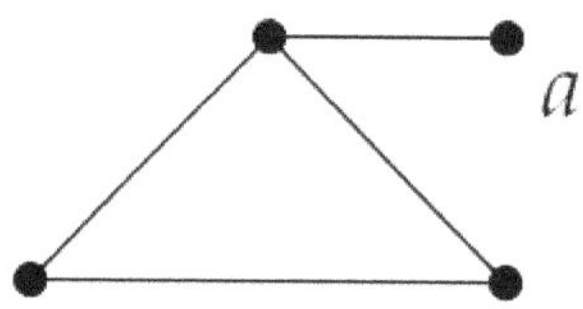

The vertex a has degree 1, and if you try to make an Euler circuit, you see that you will get stuck at the vertex. It is a dead end. That is, unless you start there. But then there is no way to return, so there is no hope of finding an Euler circuit. There is however an Euler path. It starts at the vertex a, then loops around the triangle. You will end at the vertex of degree 3.

You run into a similar problem whenever you have a vertex of any odd degree. If you start at such a vertex, you will not be able to end there (after traversing every edge exactly once). After using one edge to leave the starting vertex, you will be left with an even number of edges emanating from the vertex. Half of these could be used for returning to the vertex, the other half for leaving. So you return, then leave. Return, then leave. The only way to use up all the edges is to use the last one by leaving the vertex. On the other hand, if you have a vertex with odd degree that you do not start a path at, then you will eventually get stuck at that vertex. The path will use pairs of edges incident to the vertex to arrive and leave again.

What all this says is that if a graph has an Euler path and two vertices with odd degree, then the Euler path must start at one of the odd degree vertices and end at the other. In such a situation, every other vertex must have an even degree since we need an equal number of edges to get to those vertices as to leave them. How could we have an Euler circuit? The graph could not have any odd degree vertex as an Euler path would have to start there or end there, but not both. Thus for a graph to have an Euler circuit, all vertices must have even degree.

The converse is also true: if all the vertices of a graph have even degree, then the graph has an Euler circuit, and if there are exactly two vertices with odd degree, the graph has an Euler path. To prove this is a little tricky, but the basic idea is that you will never get stuck because there is an "outbound" edge for every "inbound" edge at every vertex. If you try to make an Euler path and miss some edges, you will always be able to "splice in" a circuit using the edges you previously missed.

Euler Paths and Circuits

- A graph has an Euler circuit if and only if the degree of every vertex is even.
- A graph has an Euler path if and only if there are at most two vertices with odd degree.

Since the bridges of Königsberg graph has all four vertices with odd degree, there is no Euler path through the graph. Thus there is no way for the townspeople to cross every bridge exactly once.

Hamilton Paths

Supposc you wanted to tour Königsberg in such a way that you visit each land mass (the two islands and both banks) exactly once. This can be done. In graph theory terms, we are asking whether there is a path which visits every vertex exactly once. Such a path is called a **Hamilton path** (or **Hamiltonian path**). We could also consider **Hamilton cycles**, which are Hamilton paths which start and stop at the same vertex.

Example 5.5.1

Determine whether the graphs below have a Hamilton path.

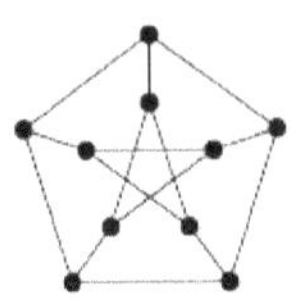

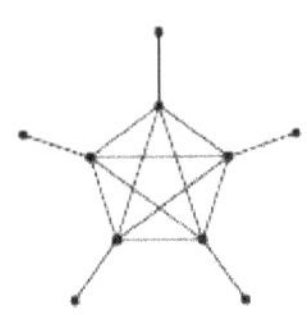

Solution

The graph on the left has a Hamilton path (many different ones, actually), as shown here:

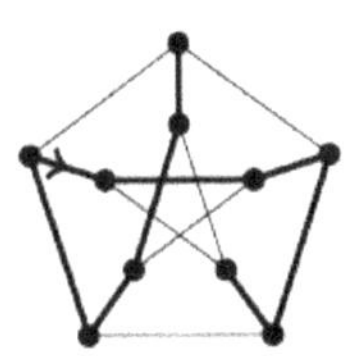

The graph on the right does not have a Hamilton path. You would need to visit each of the "outside" vertices, but as soon as you visit one, you get stuck. Note that this graph does not have an Euler path, although there are graphs with Euler paths but no Hamilton paths.

It appears that finding Hamilton paths would be easier because graphs often have more edges than vertices, so there are fewer requirements to be met. However, nobody knows whether this is true. There is no known simple test for whether a graph has a Hamilton path. For small graphs this is not a problem, but as the size of the graph grows, it gets harder and harder to check wither there is a Hamilton path. In fact, this is an example of a question which as far as we know is too difficult for computers to solve; it is an example of a problem which is NP-complete.

EXERCISES

1. **You and your friends want to tour the southwest by car. You will visit the nine states below, with the following rather odd rule: you must cross each border between neighboring states exactly once (so, for example, you must cross the Colorado-Utah border exactly once). Can you do it? If so, does it matter where you start your road trip? What fact about graph theory solves this problem?**

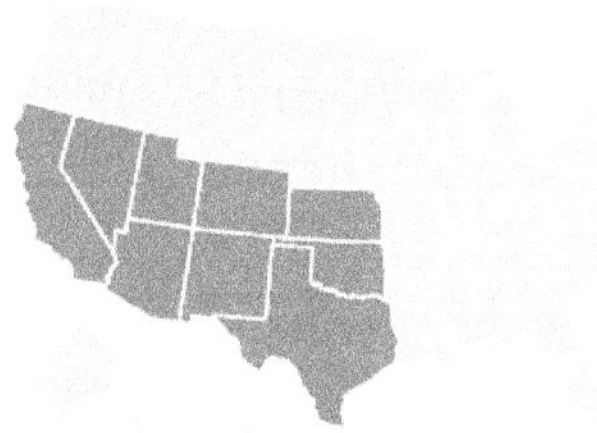

2. **Which of the following graphs contain an Euler path? Which contain an Euler circuit?**

   (a) K_4 (b) K_5. (c) $K_{5,7}$ (d) $K_{2,7}$ (e) C_7 (f) P_7

3. **Edward A. Mouse has just finished his brand new house. The floor plan is shown below:**

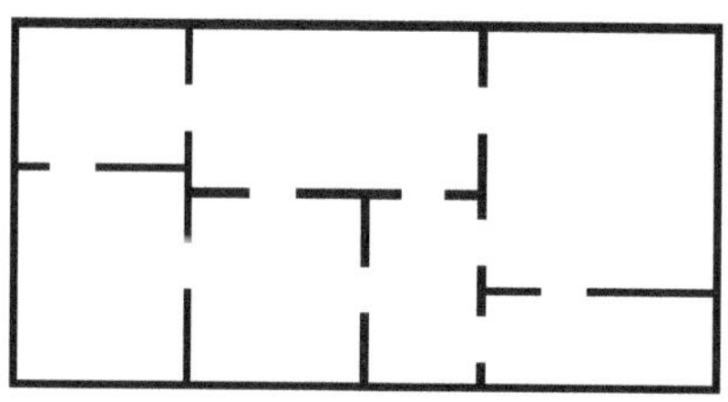

   (a) Edward wants to give a tour of his new pad to a lady-mouse- friend. Is it possible for them to walk through every doorway exactly once? If so, in which rooms must they begin and end the tour? Explain.

(b) Is it possible to tour the house visiting each room exactly once (not necessarily using every doorway)? Explain.

(c) After a few mouse-years, Edward decides to remodel. He would like to add some new doors between the rooms he has. Of course, he cannot add any doors to the exterior of the house. Is it possible for each room to have an odd number of doors? Explain.

4. **For which n does the graph K_n contain an Euler circuit? Explain.**

5. **For which m and n does the graph $K_{m,n}$ contain an Euler path? An Euler circuit? Explain.**

6. **For which n does K_n contain a Hamilton path? A Hamilton cycle? Explain.**

7. **For which m and n does the graph $K_{m,n}$ contain a Hamilton path? A Hamilton cycle? Explain.**

8. **A bridge builder has come to Königsberg and would like to add bridges so that it is possible to travel over every bridge exactly once. How many bridges must be built?**

9. **Below is a graph representing friendships between a group of students (each vertex is a student and each edge is a friendship). Is it possible for the students to sit around a round table in such a way that every student sits between two friends? What does this question have to do with paths?**

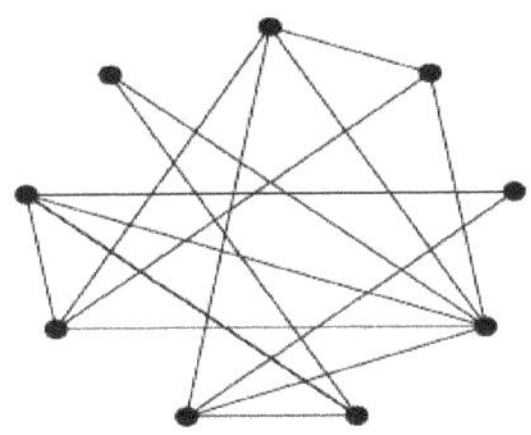

10. **On the table rest 8 dominoes, as shown below. If you were to line them up in a single row, so that any two sides touching had matching numbers, what would the sum of the two end numbers be?**

11. **Is there anything we can say about whether a graph has a Hamilton path based on the degrees of its vertices?**

(a) Suppose a graph has a Hamilton path. What is the maximum number of vertices of degree one the graph can have? Explain why your answer is correct.

(b) Find a graph which does not have a Hamilton path even though no vertex has degree one. Explain why your example works.

12. Consider the following graph:

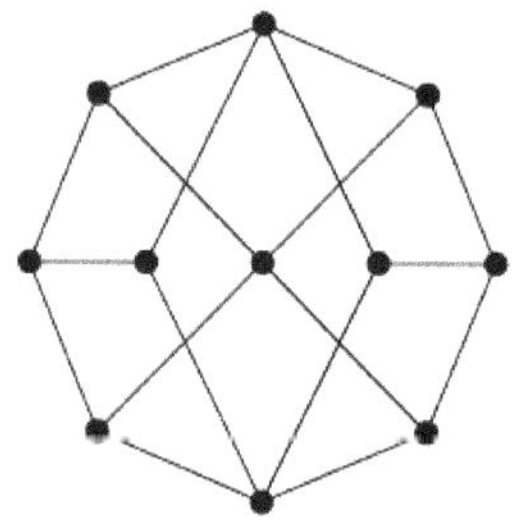

(a) Find a Hamilton path. Can your path be extended to a Hamilton cycle?

(b) Is the graph bipartite? If so, how many vertices are in each "part"?

(c) Use your answer to part (b) to prove that the graph has no Hamilton cycle.

(d) Suppose you have a bipartite graph G in which one part has at least two more vertices than the other. Prove that G does not have a Hamilton path.

5.6 MATCHING IN BIPARTITE GRAPHS

Investigate!

Given a bipartite graph, a matching is a subset of the edges for which every vertex belongs to exactly one of the edges. Our goal in this activity is to discover some criterion for when a bipartite graph has a matching.

Does the graph below contain a matching? If so, find one.

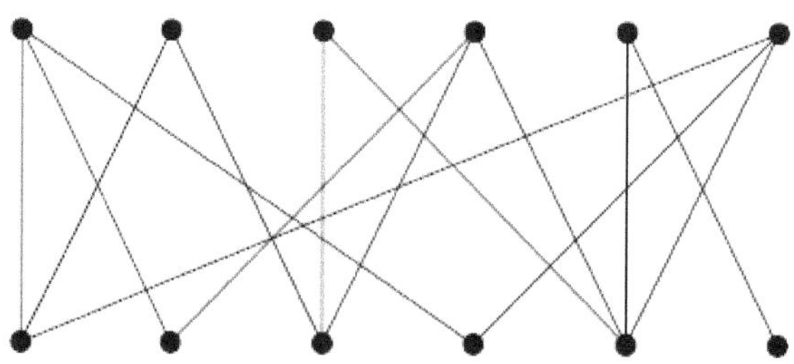

Not all bipartite graphs have matchings. Draw as many fundamentally different examples of bipartite graphs which do NOT have matchings. Your goal is to find all the possible obstructions to a graph having a perfect matching. Write down the necessary conditions for a graph to have a matching (that is, fill in the blank: If a graph has a matching, then -----). Then ask yourself whether these conditions are sufficient (is it true that if ------, then the graph has a matching?).

Attempt the Above Activity Before Proceeding

We conclude with one more example of a graph theory problem to illustrate the variety and vastness of the subject.

Suppose you have a bipartite graph G. This will consist of two sets of vertices A and B with some edges connecting some vertices of A to some vertices in B (but of course, no edges between two vertices both in A or both in B). A **matching of** A is a subset of the edges for which each vertex of A belongs to exactly one edge of the subset, and no vertex in B belongs to more than one edge in the subset. In practice we will assume that $|A| = |B|$ (the two sets have the same number of vertices) so this says that every vertex in the graph belongs to exactly one edge in the matching.

Some context might make this easier to understand. Think of the vertices in A as representing students in a class, and the vertices in B as representing presentation topics. We put an edge from a vertex $a \in A$ to a vertex $b \in B$ if student a would like to present on topic b. Of course, some students would want to present on more than one topic, so their vertex would have degree greater than 1. As the teacher, you want to assign each student their own unique topic. Thus you want to find a matching of A: you pick some subset of the edges so that each student gets matched up with exactly one topic, and no topic gets matched to two students.[11]

The question is: when does a bipartite graph contain a matching of A?

To begin to answer this question, consider what could prevent the graph from containing a matching. This will not necessarily tell us a condition when the graph does have a matching, but at least it is a start.

One way G could not have a matching is if there is a vertex in A not adjacent to any vertex in B (so having degree 0). What else? What if two students both like the same one topic, and no others? Then after assigning that one topic to the first student, there is nothing left for the second student to like, so it is very much as if the second student has degree 0. Or what if three students like only two topics between them. Again, after assigning one student a topic, we reduce this down to the previous case of two students liking only one topic. We can continue this way with more and more students.

It should be clear at this point that if there is every a group of n students who as a group like $n-1$ or fewer topics, then no matching is possible. This is true for any value of n, and any group of n students.

To make this more graph-theoretic, say you have a set $S \subseteq A$ of vertices. Define $N(S)$ to be the set of all the **neighbors** of vertices in S. That is, $N(S)$ contains all the vertices (in B) which are adjacent to at least one of the vertices in S. (In the student/topic graph, $N(S)$ is the set of topics liked by the students of S.) Our discussion above can be summarized as follows:

Matching Condition

If a bipartite graph $G = \{A, B\}$ has a matching of A, then

$$|N(S)| \geq |S|$$

for all $S \subseteq A$.

Is the converse true? Suppose G satisfies the matching condition $|N(S)| \geq |S|$ for all $S \subseteq A$ (every set of vertices has at least as many neighbors than vertices in the set). Does that mean that there is a matching?

Surprisingly, yes. The obvious necessary condition is also sufficient.12 This is a theorem first proved by Philip Hall in 1935.[13]

Theorem 5.6.1 Hall's Marriage Theorem. *Let G be a bipartite graph with sets A and B. Then G has a matching of A if and only if*

$$|N(S)| \geq |S|$$

for all $S \subseteq A$.

There are quite a few different proofs of this theorem – a quick internet search will get you started.

In addition to its application to marriage and student presentation topics, matchings have applications all over the place. We conclude with one such example.

Example 5.6.2

Suppose you deal 52 regular playing cards into 13 piles of 4 cards each. Prove that you can always select one card from each pile to get one of each of the 13 card values Ace, 2, 3, . . . , 10, Jack, Queen, and King.

Solution

Doing this directly would be difficult, but we can use the matching condition to help. Construct a graph G with 13 vertices in the set A, each representing one of the 13 card values, and 13 vertices in the set B, each representing one of the 13 piles. Draw an edge between a vertex $a \in A$ to a vertex $b \in B$ if a card with value a is in the pile b. Notice that we are just looking for a matching of A; each value needs to be found in the piles exactly once.

We will have a matching if the matching condition holds. Given any set of card values (a set $S \subseteq A$) we must show that $|N(S)| \geq |S|$. That is, the number of piles that contain those values is at least the number of different values. But what if it wasn't? Say $|S| = k$. If $|N(S)| < k$, then we would have fewer than 4k different cards in those piles (since each pile contains 4 cards). But there are 4k cards with the k different values, so at least one of these cards must be in another pile, a contradiction. Thus the matching condition holds, so there is a matching, as required.

EXERCISES

1. **Find a matching of the bipartite graphs below or explain why no matching exists.**

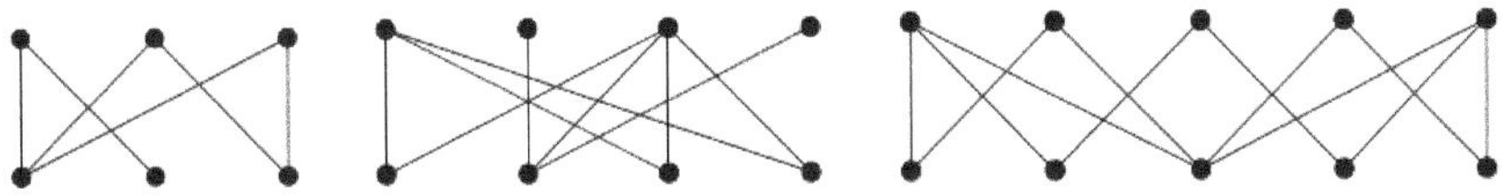

2. **A bipartite graph that doesn't have a matching might still have a partial matching. By this we mean a set of edges for which no vertex belongs to more than one edge (but possibly belongs to none). Every bipartite graph (with at least one edge) has a partial matching, so we can look for the largest partial matching in a graph.**

Your "friend" claims that she has found the largest partial matching for the graph below (her matching is in bold). She explains that no other edge can be added, because all the edges not used in her partial matching are connected to matched vertices. Is she correct?

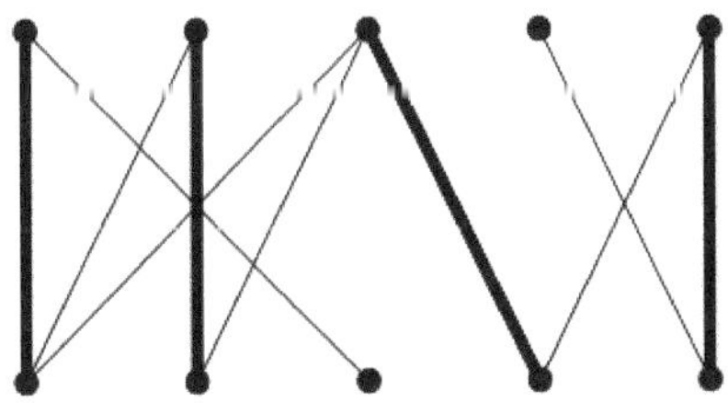

3. **One way you might check to see whether a partial matching is maximal is to construct an alternating path. This is a sequence of adjacent edges, which alternate between edges in the matching and edges not in the matching (no edge can be used more than once). If an alternating path starts and stops with an edge not in the matching, then it is called an augmenting path.**

   (a) Find the largest possible alternating path for the partial matching of your friend's graph. Is it an augmenting path? How would this help you find a larger matching?

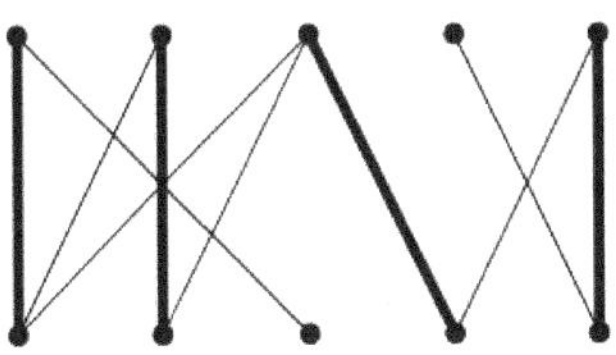

   (b) Find the largest possible alternating path for the partial matching below. Are there any augmenting paths? Is the partial matching the largest one that exists in the graph?

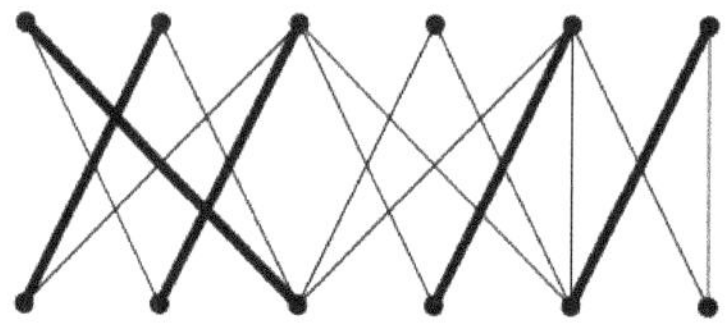

4. The two richest families in Westeros have decided to enter into an alliance by marriage. The first family has 10 sons, the second has 10 girls. The ages of the kids in the two families match up. To avoid impropriety, the families insist that each child must marry someone either their own age, or someone one position younger or older. In fact, the graph representing agreeable marriages looks like this:

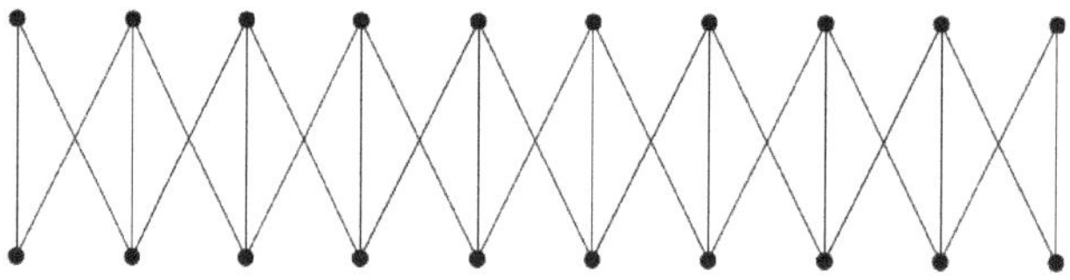

The question: how many different acceptable marriage arrangements which marry off all 20 children are possible?

   (a) How many marriage arrangements are possible if we insist that there are exactly 6 boys marry girls not their own age?
   (b) Could you generalize the previous answer to arrive at the total number of marriage arrangements?
   (c) How do you know you are correct? Try counting in a different way. Look at smaller family sizes and get a sequence.
   (d) Can you give a recurrence relation that fits the problem?

5. **We say that a set of vertices $A \subseteq V$ is a vertex cover if every edge of the graph is incident to a vertex in the cover (so a vertex cover covers the *edges*). Since V itself is a vertex cover, every graph has a vertex cover. The interesting question is about finding a minimal vertex cover, one that uses the fewest possible number of vertices.**
   (a) Suppose you had a matching of a graph. How can you use that to get a minimal vertex cover? Will your method always work?
   (b) Suppose you had a minimal vertex cover for a graph. How can you use that to get a partial matching? Will your method always work?
   (c) What is the relationship between the size of the minimal vertex cover and the size of the maximal partial matching in a graph?

6. **For many applications of matchings, it makes sense to use bipartite graphs. You might wonder, however, whether there is a way to find matchings in graphs in general.**

(a) For which n does the complete graph K_n have a matching?
(b) Prove that if a graph has a matching, then $|V|$ is even.
(c) Is the converse true? That is, do all graphs with $|V|$ even have a matching?
(d) What if we also require the matching condition? Prove or disprove: If a graph with an even number of vertices satisfies $|N(S)| \geq |S|$ for all $S \subseteq V$, then the graph has a matching.

5.7 CONCISE SUMMARY

Hopefully this chapter has given you some sense for the wide variety of graph theory topics as well as why these studies are interesting. There are many more interesting areas to consider and the list is increasing all the time; graph theory is an active area of mathematical research.

One reason graph theory is such a rich area of study is that it deals with such a fundamental concept: any pair of objects can either be related or not related. What the objects are and what "related" means varies on context, and this leads to many applications of graph theory to science and other areas of math. The objects can be countries, and two countries can be related if they share a border. The objects could be land masses which are related if there is a bridge between them. The objects could be websites which are related if there is a link from one to the other. Or we can be completely abstract: the objects are vertices which are related if their is an edge between them.

What question we ask about the graph depends on the application, but often leads to deeper, general and abstract questions worth studying in their own right. Here is a short summary of the types of questions we have considered:

Can the graph be drawn in the plane without edges crossing? If so, how many regions does this drawing divide the plane into?

Is it possible to color the vertices of the graph so that related vertices have different colors using a small number of colors? How many colors are needed?

Is it possible to trace over every edge of a graph exactly once without lifting up your pencil? What other sorts of "paths" might a graph posses?

Can you find subgraphs with certain properties? For example, when does a (bipartite) graph contain a subgraph in which all vertices are only related to one other vertex?

Not surprisingly, these questions are often related to each other. For example, the chromatic number of a graph cannot be greater than 4 when the graph is planar. Whether the graph has an Euler path depends on how many vertices each vertex is adjacent to (and whether those numbers are always even or not). Even the existence of matchings in bipartite graphs can be proved using paths.

Chapter Review

1. **Which (if any) of the graphs below are the same? Which are different? Explain.**

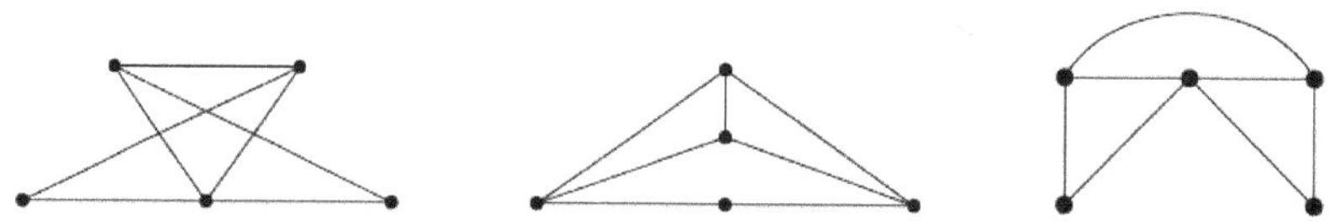

2. **Which of the graphs in the previous question contain Euler paths or circuits? Which of the graphs are planar?**

3. **Draw a graph which has an Euler circuit but is not planar.**

4. **Draw a graph which does not have an Euler path and is also not planar.**

5. **Consider the graph $G = (V, E)$ with $V = \{a, b, c, d, e, f, g\}$ and $E = \{ab, ac, af, bg, cd, ce\}$ (here we are using the shorthand for edges: ab really means $\{a, b\}$, for example).**

   (a) Is the graph G isomorphic to $G' = (V', E)$ with $V' = \{t, u, v, w, x, y, z\}$ and $E' = \{tz, uv, uy, uz, vw, vx\}$? If so, give the isomoprhism. If not, explain how you know.

   (b) Find a graph G'' with 7 vertices and 6 edges which is NOT isomorphic to G, or explain why this is not possible.

   (c) Write down the *degree sequence* for G. That is, write down the degrees of all the vertices, in decreasing order.

   (d) Find a connected graph G''' with the same degree sequence of G which is NOT isomorphic to G, or explain why this is not possible.

   (e) What kind of graph is G? Is G complete? Bipartite? A tree? A cycle? A path? A wheel?

   (f) Is G planar?

   (g) What is the chromatic number of G? Explain.

   (h) Does G have an Euler path or circuit? Explain.

6. **If a graph has 10 vertices and 10 edges and contains an Euler circuit, must it be planar? How many faces would it have?**

7. **Suppose G is a graph with n vertices, each having degree 5.**

   (a) For which values of n does this make sense?

   (b) For which values of n does the graph have an Euler path?

   (c) What is the smallest value of n for which the graph might be planar? (tricky)

8. **At a school dance, 6 girls and 4 boys take turns dancing (as couples) with each other.**
   (a) How many couples danced if every girl dances with every boy?
   (b) How many couples danced if everyone danced with everyone else (regardless of gender)?
   (c) Explain what graphs can be used to represent these situations.
9. **Among a group of n people, is it possible for everyone to be friends with an odd number of people in the group? If so, what can you say about n?**
10. **Your friend has challenged you to create a convex polyhedron containing 9 triangles and 6 pentagons.**
    (a) Is it possible to build such a polyhedron using only these shapes? Explain.
    (b) You decide to also include one heptagon (seven-sided polygon). How many vertices does your new convex polyhedron contain?
    (c) Assuming you are successful in building your new 16-faced polyhedron, could every vertex be the joining of the same number of faces? Could each vertex join either 3 or 4 faces? If so, how many of each type of vertex would there be?
11. **Is there a convex polyhedron which requires 5 colors to properly color the vertices of the polyhedron? Explain.**
12. **How many edges does the graph $K_{n,n}$ have? For which values of n does the graph contain an Euler circuit? For which values of n is the graph planar?**
13. **The graph G has 6 vertices with degrees 1, 2, 2, 3, 3, 5. How many edges does G have? If G was planar how many faces would it have? Does G have an Euler path?**
14. **What is the smallest number of colors you need to properly color the vertices of K_7. Can you say whether K_7 is planar based on your answer?**
15. **What is the smallest number of colors you need to properly color the vertices of $K_{3,4}$? Can you say whether $K_{3,4}$ is planar based on your answer?**
16. **Prove that $K_{3,4}$ is not planar. Do this using Euler's formula, not just by appealing to the fact that $K_{3,3}$ is not planar.**
17. **A dodecahedron is a regular convex polyhedron made up of 12 regular pentagons.**
    (a) Suppose you color each pentagon with one of three colors. Prove that there must be two adjacent pentagons colored identically.
    (b) What if you use four colors?
    (c) What if instead of a dodecahedron you colored the faces of a cube?

18. Decide whether the following statements are true or false. Prove your answers.

(a) If two graph G_1 and G_2 have the same chromatic number, then they are isomorphic.

(b) If two graphs G_1 and G_2 have the same number of vertices and edges and have the same chromatic number, then they are isomorphic.

(c) If two graphs are isomorphic, then they have the same chromatic number.

19. If a planar graph G with 7 vertices divides the plane into 8 regions, how many edges must G have?

20. Consider the graph below:

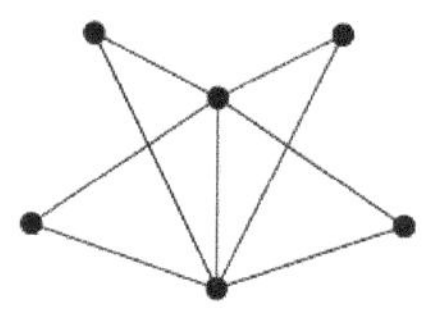

(a) Does the graph have an Euler path or circuit? Explain.

(b) Is the graph planar? Explain.

(c) Is the graph bipartite? Complete? Complete bipartite?

(d) What is the chromatic number of the graph.

21. For each part below, say whether the statement is true or false. Explain why the true statements are true, and give counterexamples for the false statements.

(a) Every bipartite graph is planar.

(b) Every bipartite graph has chromatic number 2.

(c) Every bipartite graph has an Euler path.

(d) Every vertex of a bipartite graph has even degree.

(e) A graph is bipartite if and only if the sum of the degrees of all the vertices is even.

22. Consider the statement "If a graph is planar, then it has an Euler path."

(a) Write the converse of the statement.

(b) Write the contrapositive of the statement.

(c) Write the negation of the statement.

(d) Is it possible for the contrapositive to be false? If it was, what would that tell you?

(e) Is the original statement true or false? Prove your answer.

(f) Is the converse of the statement true or false? Prove your answer.

23. Let G be a connected graph with v vertices and e edges. Use mathematical induction to prove that if G contains exactly one cycle (among other edges and vertices), then $v = e$.

Note: this is asking you to prove a special case of Euler's formula for planar graphs, so do not use that formula in your proof.

FOOTNOTES

1. This is not unlike geometry, where we might have more than one copy of a particular triangle. There instead of *isomorphic* we say *congruent.*
2. A *lemma* is a mathematical statement that is primarily of importance in that it is used to establish other results.
3. A **proposition** is a general statement in mathematics, similar to a theorem, although generally of lesser importance.
4. Sometimes this is stated as "a tree is an acyclic connected graph;" "acyclic" is just a fancy word for "containing no cycles."
5. A corollary is another sort of provable statement, like a proposition or theorem, but one that follows direction from another already established statement, or its proof.
6. If you add the smallest edge adjacent to edges you have already added, you are doing *Prim's algorithm*. If you add the smallest edge in the entire graph, you are following *Kruskal's algorithm.*
7. 7An alternative definition for convex is that the internal angle formed by any two faces must be less than 180 deg.
8. Notice that you can tile the plane with hexagons. This is an infinite planar graph; each vertex has degree 3. These infinitely many hexagons correspond to the limit as $f \to \infty$ 1 to make $k = 3$.
9. There are special classes of graphs which can be proved to be perfect. One such class is the set of **chordal** graphs, which have the property that every cycle in the graph contains a **chord**—an edge between two vertices in of the cycle which are not adjacent in the cycle.
10. Note: what we are calling a *matching* is sometimes called a *perfect matching* or *complete matching*. This is because in it interesting to look at non-perfect matchings as well. We will call those *partial* matchings.
11. The standard example for matchings used to be the *marriage problem* in which A consisted of the men in the town, B the women, and an edge represented a marriage that was agreeable to both parties. A matching then represented a way for the town elders to marry off everyone in the

town, no polygamy allowed. We have chosen a more progressive context for the sake of political correctness.

12. This happens often in graph theory. If you can avoid the obvious counterexamples, you often get what you want.
13. There is also an infinite version of the theorem which was proved by Marshal Hall, Jr. The name is a coincidence though as the two Halls are not related.

CHAPTER-6

ADDITIONAL TOPICS IN DISCRETE MATHEMATICS

INTRODUCTION

Additional topics in discrete mathematics encompass a diverse array of advanced concepts that build upon the fundamental principles of the discipline. While discrete mathematics traditionally covers foundational topics such as set theory, logic, and combinatorics, these additional topics extend into more specialized areas. This includes advanced graph theory, which explores complex networks and connectivity patterns, as well as cryptography, which involves secure communication and data protection algorithms. Other areas like game theory analyze strategic decision-making, while number theory delves into the properties of integers and their relationships. Each of these topics plays a crucial role in both theoretical research and practical applications across fields such as computer science, engineering, economics, and beyond. This exploration will delve into the intricacies of these additional topics, highlighting their significance, applications, and the interconnectedness that enriches the study of discrete mathematics.

6.1 GENERATING FUNCTIONS

There is an extremely powerful tool in discrete mathematics used to manipulate sequences called the generating function. The idea is this: instead of an infinite sequence (for example: 2, 3, 5, 8, 12, . . .) we look at a single function which encodes the sequence. But not a function which gives the nth term as output. Instead, a function whose power series (like from calculus) "displays" the terms of the sequence. So for example, we would look at the power series 2 + 3x + 5x2 + 8x3 + 12x4 + ... which displays the sequence 2, 3, 5, 8, 12, . . . as coefficients.

An infinite power series is simply an infinite sum of terms of the form cnxn were cn is some constant. So we might write a power series like this:

$$\sum_{k=0}^{\infty} c_k x^k.$$

or expanded like this

$$c_0 + c_1 x + c_2 x^2 + c_3 x^3 + c_4 x^4 + c_5 x^5 + \cdots.$$

When viewed in the context of generating functions, we call such a power series a generating series. The generating series generates the sequence

$$c_0, c_1, c_2, c_3, c_4, c_5, \ldots.$$

In other words, the sequence generated by a generating series is simply the sequence of coefficients of the infinite polynomial.

Example 6.1.1

What sequence is represented by the generating series $3 + 8x^2 + x^3 + \frac{x^5}{7} + 100x^6 + \cdots$?

Solution. We just read off the coefficients of each xn term. So a0 = 3 since the coefficient of x0 is 3 (x0 = 1 so this is the constant term). What is a1? It is NOT 8, since 8 is the coefficient of x2, so 8 is the term a2 of the sequence. To find a1 we need to look for the coefficient of x1 which in this case is 0. So a1 = 0. Continuing, we have a2 = 8, a3 = 1, a4 = 0, and $a_5 = \frac{1}{7}$. So we have the sequence

$$3, 0, 8, 1, 0, \frac{1}{7}, 100, \ldots.$$

Note that when discussing generating functions, we always start our sequence with a0.

3, 0, 8, 1, 0, 1 , 100, ...

Now you might very naturally ask why we would do such a thing. One reason is that encoding a sequence with a power series helps us keep track of which term is which in the sequence. For example, if we write the sequence 1, 3, 4, 6, 9, . . . , 24, 41, . . . it is impossible to determine which term 24 is (even if we agreed that the first term was supposed to be a0). However, if we wrote the generating series instead, we would have 1 + 3x + 4x2 + 6x3 + 9x4 + ... + 24x17 + 41x18 + ... Now it is clear that 24 is the 17th term of the sequence (that is, a17 = 24). Of course to get this benefit we could have displayed our sequence in any number of ways, perhaps $\boxed{1}_0\boxed{3}_1\boxed{4}_2\boxed{6}_3\boxed{9}_4\cdots\boxed{24}_{17}\boxed{41}_{18}\cdots$,but we do not do this. The reason is that the generating series looks like an ordinary power series (although we are interpreting it differently) so we can do things with it that we ordinarily do with power series such as write down what it converges to.

For example, from calculus we know that the power series $1+x+\frac{x^2}{2}+\frac{x^3}{6}+\frac{x^4}{24}+\cdots+\frac{x^n}{n!}+\cdots$ converges to the function ex . So we can use ex as a way of talking about the sequence of coefficients of the power series for ex . When we write down a nice compact function which has an infinite power series that we view as a generating series, then we call that function a generating function. In this example, we would say

$$1,1,\frac{1}{2},\frac{1}{6},\frac{1}{24},\ldots,\frac{1}{n!},\ldots \text{ has generating function } e^x.$$

Building Generating Functions

The ex example is very specific. We have a rather odd sequence, and the only reason we know its generating function is because we happen to know the Taylor series for ex . Our goal now is to gather some tools to build the generating function of a particular given sequence.

Let's see what the generating functions are for some very simple sequences. The simplest of all: 1,1, 1, 1, 1, ... What does the generating series look like? It is simply 1 + x + x2 + x3 + x4 + ... Now, can we find a closed formula for this power series? Yes! This particular series is really just a geometric series with common ratio x. So if we use our "multiply, shift and subtract" technique from Section 2.2, we have

$$\begin{array}{rl} S = & 1+x+x^2+x^3+\cdots \\ - \quad xS = & \quad\; x+x^2+x^3+x^4+\cdots \\ \hline (1-x)S = & 1 \end{array}$$

Therefore we see that

$$1+x+x^2+x^3\cdots=\frac{1}{1-x}.$$

You might remember from calculus that this is only true on the interval of convergence for the power series, in this case when |x| < 1. That is true for us, but we don't care. We are never going to plug anything in for x, so as long as there is some value of x for which the generating function and generating series agree, we are happy. And in this case we are happy.

1, 1, 1, ...

The generating function for 1, 1, 1, 1, 1, 1, . . . is $\frac{1}{1-x}$

Let's use this basic generating function to find generating functions for more sequences. What if we replace x by -x. We get

$$\frac{1}{1+x} = 1 - x + x^2 - x^3 + \cdots$$

which generates 1, -1, 1, -1, ...

If we replace x by 3x we get

$$\frac{1}{1-3x} = 1 + 3x + 9x^2 + 27x^3 + \cdots$$

which generates 1, 3, 9, 27, ...

By replacing the in $\frac{1}{1-x}$ we can get generating functions for a variety of sequences, but not all. For example, you cannot plug in anything for x to get the generating function for 2, 2, 2, 2, ... However, we are not lost yet. Notice that each term of 2, 2, 2, 2, ... is the result of multiplying the terms of 1, 1, 1, 1, . . . by the constant 2. So multiply the generating function by 2 as well.

$$\frac{2}{1-x} = 2 + 2x + 2x^2 + 2x^3 + \cdots$$

which generates 2, 2, 2, 2, ...

Similarly, to find the generating function for the sequence 3, 9, 27, 81, . . .,

we note that this sequence is the result of multiplying each term of 1, 3, 9, 27, . . . by 3. Since we have the generating function for 1, 3, 9, 27, . . . we can say

$$\frac{3}{1-3x} = 3 \cdot 1 + 3 \cdot 3x + 3 \cdot 9x^2 + 3 \cdot 27x^3 + \cdots$$

which generates 3, 9, 27, 81, ...

What about the sequence 2, 4, 10, 28, 82, . . .? Here the terms are always 1 more than powers of 3. That is, we have added the sequences 1, 1, 1, 1, . . . and 1, 3, 9, 27, . . . term by term. Therefore we can get a generating function by adding the respective generating functions:

$$\begin{aligned} 2 + 4x + 10x^2 + 28x^3 + \cdots &= (1+1) + (1+3)x + (1+9)x^2 + (1+27)x^3 + \cdots \\ &= 1 + x + x^2 + x^3 + \cdots + 1 + 3x + 9x^2 + 27x^3 + \cdots \\ &= \frac{1}{1-x} + \frac{1}{1-3x} \end{aligned}$$

The fun does not stop there: if we replace x in our original generating function by x2 we get

$$\frac{1}{1-x^2} = 1 + x^2 + x^4 + x^6 \cdots$$

which generates 1, 0, 1, 0, 1, 0, ...

How could we get 0, 1, 0, 1, 0, 1, ? Start with the previous sequence and shift it over by 1. But how do you do this? To see how shifting works, let's first try to get the generating function for the sequence 0, 1, 3, 9, 27, ... We know that $\frac{1}{1-3x} = 1 + 3x + 9x^2 + 27x^3 + \cdots$. To get the zero out front, we need the generating series to look like x + 3x2+ 9x3+ 27x4+ ··· (so there is no constant term). Multiplying by x has this effect. So the generating function for 0, 1, 3, 9, 27, . . . is $\frac{x}{1-3x}$. This will also work to get the generating function for 0, 1, 0, 1, 0, 1, . . .:

$$\frac{x}{1-x^2} = x + x^3 + x^5 + \cdots$$

which generates 0, 1, 0, 1, 0, 1 . . . .

What if we add the sequences 1, 0, 1, 0, 1, 0, . . . and 0, 1, 0, 1, 0, 1, . . . term by term? We should get 1, 1, 1, 1, 1, 1 What happens when we add the generating functions? It works (try it)!

$$\frac{1}{1-x^2} + \frac{x}{1-x^2} = \frac{1}{1-x}.$$

Here's a sneaky one: what happens if you take the derivative of $\frac{1}{1-x}$?

We get $\frac{1}{(1-x)^2}$. On the other hand, if we differentiate term by term in the power series, we get (1 + x + x2 + x3 +....)' = 1 + 2x + 3x2 + 4x3 +..... which is the generating series for 1, 2, 3, 4, ... This says

1, 2, 3, ...

The generating function for 1, 2, 3, 4, 5, . . . is $\frac{1}{(1-x)^2}$.

Take a second derivative: $\frac{2}{(1-x)^3} = 2 + 6x + 12x^2 + 20x^3 + \cdots$.

So $\frac{1}{(1-x)^3} = 1 + 3x + 6x^2 + 10x^3 + \cdots$ is a generating function for the triangular numbers, 1, 3, 6, 10 . . . (although here we have a0 = 1 while T0 = 0 usually).

Differencing

We have seen how to find generating functions from $\frac{1}{1-x}$ using multiplication (by a constant or by x), substitution, addition, and differentiation.

To use each of these, you must notice a way to transform the sequence 1, 1, 1, 1, 1 . . . into your desired sequence. This is not always easy. It is also not really the way we have analyzed sequences. One thing we have considered often is the sequence of differences between terms of a sequence. This will turn out to be helpful in finding generating functions as well. The sequence of differences is often simpler than the original sequence. So if we know a generating function for the differences, we would like to use this to find a generating function for the original sequence.

For example, consider the sequence 2, 4, 10, 28, 82, ... How could we move to the sequence of first differences: 2, 6, 18, 54,........ ? We want to subtract 2 from the 4, 4 from the 10, 10 from the 28, and so on. So if we subtract (term by term) the sequence 0, 2, 4, 10, 28, ... from 2, 4, 10, 28 .... , we will be set. We can get the generating function for 0, 2, 4, 10, 28, . . . from the generating function for 2, 4, 10,

28 by multiplying by x. Use A to represent the generating function for 2, 4, 10, 28, 82, ... Then:

$$\begin{aligned} A &= 2 + 4x + 10x^2 + 28x^3 + 82x^4 + \cdots \\ - \quad xA &= 0 + 2x + 4x^2 + 10x^3 + 28x^4 + 82x^5 + \cdots \\ \hline (1-x)A &= 2 + 2x + 6x^2 + 18x^3 + 54x^4 + \cdots \end{aligned}$$

While we don't get exactly the sequence of differences, we do get something close. In this particular case, we already know the generating function A (we found it in the previous section) but most of the time we will use this differencing technique to find A: if we have the generating function for the sequence of differences, we can then solve for A.

Example 5.1.2

Find a generating function for 1, 3, 5, 7, 9, ...

Solution. Notice that the sequence of differences is constant. We know how to find the generating function for any constant sequence. So denote the generating function for 1, 3, 5, 7, 9, ... by A. We have

$$\begin{aligned} A &= 1 + 3x + 5x^2 + 7x^3 + 9x^4 + \cdots \\ - \quad xA &= 0 + x + 3x^2 + 5x^3 + 7x^4 + 9x^5 + \cdots \\ \hline (1-x)A &= 1 + 2x + 2x^2 + 2x^3 + 2x^4 + \cdots \end{aligned}$$

We know that $2x + 2x^2 + 2x^3 + 2x^4 + \cdots = \frac{2x}{1-x}$. Thus

$$(1-x)A = 1 + \frac{2x}{1-x}.$$

Now solve for A:

$$A = \frac{1}{1-x} + \frac{2x}{(1-x)^2} = \frac{1+x}{(1-x)^2}.$$

Does this makes sense? Before we simplified the two fractions into one, we were adding the generating function for the sequence 1, 1, 1, 1, . . . to the generating function for the sequence 0, 2, 4, 6, 8, 10, . . . (remember $\frac{1}{(1-x)^2}$ generates 1, 2, 3, 4, 5, . . ., multiplying by 2x shifts it over, putting the zero out front, and doubles each term). If we add these term by term, we get the correct sequence 1, 3, 5, 7, 9, ...

Now that we have a generating function for the odd numbers, we can use that to find the generating function for the squares:

Example 6.1.3

Find the generating function for 1, 4, 9, 16, ... Note we take 1 = a0.

Solution

Again we call the generating function for the sequence A. Using differencing:

$$\begin{array}{rl} A &= 1 + 4x + 9x^2 + 16x^3 + \cdots \\ - \quad xA &= 0 + x + 4x^2 + 9x^3 + 16x^4 + \cdots \\ \hline (1 - x)A &= 1 + 3x + 5x^2 + 7x^3 + \cdots \end{array}$$

Since $1 + 3x + 5x^2 + 7x^3 + \cdots = \frac{1+x}{(1-x)^2}$ we have $A = \frac{1+x}{(1-x)^3}$.

In each of the examples above, we found the difference between consecutive terms which gave us a sequence of differences for which we knew a generating function. We can generalize this to more complicated relationships between terms of the sequence. For example, if we know that the sequence satisfies the recurrence relation an = 3an−1 2an−2? In other words, if we take a term of the sequence and subtract 3 times the previous term and then add 2 times the term before that, we get 0 (since an−3an−1 + 2an−2 = 0). That will hold for all but the first two terms of the sequence. So after the first two terms, the sequence of results of these calculations would be a sequence of 0's, for which we definitely know a generating function.

Example 5.1.4

The sequence 1, 3, 7, 15, 31, 63, . . . satisfies the recurrence relation an = 3an−1 - 2an−2. Find the generating function for the sequence.

Solution. Call the generating function for the sequence A. We have

$$\begin{array}{rl} A &= 1 + 3x + 7x^2 + 15x^3 + 31x^4 + \cdots + a_n x^n + \cdots \\ -3xA &= 0 - 3x - 9x^2 - 21x^3 - 45x^4 - \cdots - 3a_{n-1}x^n - \cdots \\ + \quad 2x^2A &= 0 + 0x + 2x^2 + 6x^3 + 14x^4 + \cdots + 2a_{n-2}x^n + \cdots \\ \hline (1 - 3x + 2x^2)A &= 1 \end{array}$$

We multiplied A by −3x which shifts every term over one spot and multiplies them by −3. On the third line, we multiplied A by 2x2, which shifted every term over two spots and multiplied them by 2. When we add up the corresponding terms, we are taking each term, subtracting 3 times the previous term, and adding 2 times the term before that. This will happen for each term after a1 because an−3an−1 + 2an−2 = 0. In general, we might have two terms from the beginning of the generating series, although in this case the second term happens to be 0 as well.

Now we just need to solve for A:

$$A = \frac{1}{1 - 3x + 2x^2}.$$

Multiplication and Partial Sums

What happens to the sequences when you multiply two generating func- tions? Let's see: A = a0 + a1x + a2x2 + · · · · · · and B = b0 + b1x + b2x2 + ... To multiply A and B, we need to do a lot of distributing (infinite FOIL?) but keep in mind we will group like terms and only need to write down the first few terms to see the pattern. The constant term is a0b0. The coefficient of x is a0b1 + a1b0. And so on. We get:

$$AB = a_0b_0+(a_0b_1+a_1b_0)x+(a_0b_2+a_1b_1+a_2b_0)x^2+(a_0b_3+a_1b_2+a_2b_1+a_3b_0)x^3+\cdots.$$

"Multiply" the sequence 1, 2, 3, 4, . . . by the sequence 1, 2, 4, 8, 16, ...

Solution. The new constant term is just 1 · 1. The next term will be 1 · 2 + 2 · 1 = 4. The next term: 1 · 4 + 2 · 2 + 3 · 1 = 11. One more: 1 · 8 + 2 · 4 + 3 · 2 + 4 · 1 = 26. The resulting sequence is

1, 4, 11, 26, 57, ...

Since the generating function for 1, 2, 3, 4, . . . is $\frac{1}{(1-x)^2}$ and the generating function for 1, 2, 4, 8, 16 , . . . is $\frac{1}{1-2x}$, we have that the generating function for 1, 4, 11, 26, 57, . . . is $\frac{1}{(1-x)^2(1-2x)}$

Consider the special case when you multiply a sequence by 1, 1, 1, ... For example, multiply 1, 1, 1, . . . by 1, 2, 3, 4, 5 . . .. The first term is 1. 1 = 1. Then 1.2 + 1.1 = 3. Then 1. 3 + 1. 2 + 1.1 = 6. The next term will be 10. We are getting the triangular numbers. More precisely, we get the sequence of partial sums of 1, 2, 3, 4, 5, · · · · · In terms of generating functions, we take $\frac{1}{1-x}$ (generating 1, 1, 1, 1, 1 . . .) and multiply it by $\frac{1}{(1-x)^2}$ (generating 1, 2, 3, 4, 5, . . .) and this give $\frac{1}{(1-x)^3}$. This should not be a surprise as we found the same generating function for the triangular numbers earlier.

The point is, if you need to find a generating function for the sum of the first n terms of a particular sequence, and you know the generating function for that sequence, you can multiply it by $\frac{1}{1-x}$. To go back from the sequence of partial sums to the original sequence, you look at the sequence of differences. When you get the sequence of differences you end up multiplying by 1 - x, or equivalently, dividing by $\frac{1}{1-x}$. Multiplying by $\frac{1}{1-x}$ gives partial sums, dividing by $\frac{1}{1-x}$ gives differences.

Solving Recurrence Relations with Generating Functions

We conclude with an example of one of the many reasons studying generating functions is helpful. We can use generating functions to solve recurrence relations.

Example 6.1.6

Solve the recurrence relation an = 3an−1 2an−2 with initial conditions a0 = 1 and a1 = 3.

Solution

We saw in an example above that this recurrence relation gives the sequence 1, 3, 7, 15, 31, 63, . . . which has generating function $\frac{1}{1-3x+2x^2}$. We did this by calling the generating function A and then computing A–3xA + 2x2A which was just 1, since every other term canceled out.

But how does knowing the generating function help us? First, break up the generating function into two simpler ones. For this, we can use partial fraction decomposition. Start by factoring the denominator:

$$\frac{1}{1-3x+2x^2} = \frac{1}{(1-x)(1-2x)}.$$

Partial fraction decomposition tells us that we can write this faction as the sum of two fractions (we decompose the given fraction):

$\frac{1}{(1-x)(1-2x)} = \frac{a}{1-x} + \frac{b}{1-2x}$ for some constants a and b.

To find a and b we add the two decomposed fractions using a common denominator. This gives

$$\frac{1}{(1-x)(1-2x)} = \frac{a(1-2x)+b(1-x)}{(1-x)(1-2x)}.$$

so

$$1 = a(1-2x) + b(1-x).$$

This must be true for all values of x. If x = 1, then the equation becomes 1 = – a so a = –1. When $x = \frac{1}{2}$ we get 1 = b/2 so b = 2.

This tells us that we can decompose the fraction like this:

$$\frac{1}{(1-x)(1-2x)} = \frac{-1}{1-x} + \frac{2}{1-2x}.$$

This completes the partial fraction decomposition. Notice that these two fractions are generating functions we know. In fact, we should be able to expand each of them.

$$\frac{-1}{1-x} = -1-x-x^2-x^3-x^4-\cdots$$

which generates –1, –1, –1, –1, –1, ...

$$\frac{2}{1-2x} = 2+4x+8x^2+16x^3+32x^4+\cdots$$

which generates 2, 4, 8, 16, 32, ...

We can give a closed formula for the nth term of each of these sequences. The first is just an = –1. The second is an = 2n+1. The sequence we are interested in is just the sum of these, so the solution to the recurrence relation is an = 2n+1 - 1.

We can now add generating functions to our list of methods for solving recurrence relations.

Exercises

1. **Find the generating function for each of the following sequences by relating them back to a sequence with known generating function.**

   (a) 4, 4, 4, 4, 4, ...

   (b) 2, 4, 6, 8, 10, ...

   (c) 0, 0, 0, 2, 4, 6, 8, 10, ...

   (d) 1, 5, 25, 125, ...

   (e) 1, −3, 9, −27, 81, ...

   (f) 1, 0, 5, 0, 25, 0, 125, 0, ...

   (g) 0, 1, 0, 0, 2, 0, 0, 3, 0, 0, 4, 0, 0, 5, ...

2. **Find the sequence generated by the following generating functions:**

   (a) $\frac{4x}{1-x}$. (b) $\frac{1}{1-4x}$.

   (c) $\frac{x}{1+x}$. (d) $\frac{3x}{(1+x)^2}$.

   (e) $\frac{1+x+x^2}{(1-x)^2}$ (Hint: multiplication).

3. **Show how you can get the generating function for the triangular numbers in three different ways:**

   (a) Take two derivatives of the generating function for 1, 1, 1, 1, 1, . . .

   (b) Use differencing.

   (c) Multiply two known generating functions.

4. **Use differencing to find the generating function for 4, 5, 7, 10, 14, 19, 25, . .**

5. **Find a generating function for the sequence with recurrence relation an = 3an−1 - an−2 with initial terms a0 = 1 and a1 = 5.**

6. **Use the recurrence relation for the Fibonacci numbers to find the generating function for the Fibonacci sequence.**

7. **Use multiplication to find the generating function for the sequence of partial sums of Fibonacci numbers, S0, S1, S2,......... where S0 = F0, S1 = F0 + F1, S2 = F0 + F1 + F2, S3 = F0 + F1 + F2 + F3 and so on.**

8. **Find the generating function for the sequence with closed formula an = 2(5n) + 7(−3)n.**

9. **Find a closed formula for the nth term of the sequence with generating function $\frac{3x}{1-4x} + \frac{1}{1-x}$.**

10. **Find a7 for the sequence with generating function $\frac{2}{(1-x)^2} \cdot \frac{x}{1-x-x^2}$.**

11. **Explain how we know that $\frac{1}{(1-x)^2}$ is the generating function for 1, 2, 3, 4, . .**

12. **Starting with the generating function for 1, 2, 3, 4, . . ., find a generating function for each of the following sequences.**

    (a) 1, 0, 2, 0, 3, 0, 4, ...

    (b) 1, -2, 3, -4, 5, -6, ...

    (c) 0, 3, 6, 9, 12, 15, 18, ...

    (d) 0, 3, 9, 18, 30, 45, 63, (Hint: relate this sequence to the previous one.)

13 **You may assume that 1, 1, 2, 3, 5, 8, ... has generating function $\frac{1}{1-x-x^2}$ (because it does). Use this fact to find the sequence generated by each of the following generating functions.**

    (a) $\frac{x^2}{1-x-x^2}$.   (b) $\frac{1}{1-x^2-x^4}$.

    (c) $\frac{1}{1-3x-9x^2}$.   (d) $\frac{1}{(1-x-x^2)(1-x)}$.

14. **Find the generating function for the sequence 1, -2, 4, -8, 16, ...**

15. **Find the generating function for the sequence 1, 1, 1, 2, 3, 4, 5, 6, ...**

16. **Suppose A is the generating function for the sequence 3, 5, 9, 15, 23, 33, ...**

    (a) Find a generating function (in terms of A) for the sequence of differences between terms.

    (b) Write the sequence of differences between terms and find a generating function for it (without referencing A).

    (c) Use your answers to parts (a) and (b) to find the generating function for the original sequence.

6.2 INTRODUCTION TO NUMBER THEORY

We have used the natural numbers to solve problems. This was the right set of numbers to work with in discrete mathematics because we always dealt with a whole number of things. The natural numbers have been a tool. Let's take a moment now to inspect that tool. What mathematical discoveries can we make about the natural numbers themselves?

This is the main question of number theory: a huge, ancient, complex, and above all, beautiful branch of mathematics. Historically, number theory was known as the Queen of Mathematics and was very much a branch of pure mathematics, studied for its own sake instead of as a means to understanding real world applications. This has changed in recent years however, as applications of number theory have been unearthed. Probably the most well known example of this is RSA cryptography, one of the methods used in encrypt data on the internet. It is number theory that makes this possible.

- What sorts of questions belong to the realm of number theory? Here is a motivating example. Recall in our study of induction, we asked:
- Which amounts of postage can be made exactly using just 5-cent and 8-cent stamps?

We were able to prove that any amount greater than 27 cents could be made. You might wonder what would happen if we changed the denomination of the stamps. What if we instead had 4- and 9-cent stamps? Would there be some amount after which all amounts would be possible? Well, again, we could replace two 4-cent stamps with a 9-cent stamp, or three 9-cent stamps with seven 4-cent stamps. In each case we can create one more cent of postage. Using this as the inductive case would allow us to prove that any amount of postage greater than 23 cents can be made.

What if we had 2-cent and 4-cent stamps. Here it looks less promising. If we take some number of 2-cent stamps and some number of 4-cent stamps, what can we say about the total? Could it ever be odd? Doesn't look like it.

Why does 5 and 8 work, 4 and 9 work, but 2 and 4 not work? What is it about these numbers? If I gave you a pair of numbers, could you tell me right away if they would work or not? We will answer these questions, and more, after first investigating some simpler properties of numbers themselves.

Divisibility

It is easy to add and multiply natural numbers. If we extend our focus to all integers, then subtraction is also easy (we need the negative numbers so we can subtract any number from any other number, even larger from smaller). Division is the first operation that presents a challenge. If we wanted to extend our set of numbers so any division would be possible (maybe excluding division by 0) we would need to look at the rational numbers (the set of all numbers which can be written as fractions). This would be going too far, so we will refuse this option.

In fact, it is a good thing that not every number can be divided by other numbers. This helps us understand the structure of the natural numbers and opens the door to many interesting questions and applications.

If given numbers a and b, it is possible that $a \div b$ gives a whole number. In this case, we say that b divides a, in symbols, we write $b|a$. If this holds, then b is a divisor or factor of a, and a is a multiple of b. In other words, if b | a, then a = bk for some integer k (this is saying a is some multiple of b).

The Divisibility Relation

Given integers m and n, we say "m divides n" and write

$$m|n$$

provided n ÷ m is an integer. Thus the following assertions mean the same thing:

$$m \mid n$$

1. n = mk for some integer k
2. m is a factor (or divisor) of n
3. n is a multiple of m.

Notice that m|n is a statement. It is either true or false. On the other hand, n ÷ m or n/m is some number. If we want to claim that n m is not an integer, so m does not divide n, then we can write m/n.

Example 6.2.1

Decide whether each of the statements below are true or false.

1. 4 | 20
2. 20 | 4
3. 0 | 5
4. 5 | 0
5. 7 | 7
6. 1 | 37
7. -3 | 12
8. 8 | 12
9. 1642 | 136299

Solution

1. True. 4 "goes into" 20 five times without remainder. In other words, 20 ÷ 4 = 5, an integer. We could also justify this by saying that 20 is a multiple of 4: 20 = 4 · 5.
2. False. While 20 is a multiple of 4, it is false that 4 is a multiple of 20.
3. False. 5 ÷ 0 is not even defined, let alone an integer.

4. True. In fact, $x \mid 0$ is true for all x. This is because 0 is a multiple of every number: $0 = x \cdot 0$.
5. True. In fact, $x \mid x$ is true for all x.
6. True. 1 divides every number (other than 0).
7. True. Negative numbers work just fine for the divisibility relation. Here $12 = -3 \cdot 4$. It is also true that $3 \mid -12$ and that $-3 \mid -12$.
8. False. Both 8 and 12 are divisible by 4, but this does not mean that 12 is divisible by 8.
9. False. See below.

This last example raises a question: how might one decide whether $m|n$? Of course, if you had a trusted calculator, you could ask it for the value of $n \div m$. If it spits out anything other than an integer, you know $m \nmid n$. This seems a little like cheating though: we don't have division, so should we really use division to check divisibility?

While we don't really know how to divide, we do know how to multiply. We might try multiplying m by larger and larger numbers until we get close to n. How close? Well, we want to be sure that if we multiply m by the next larger integer, we go over n.

For example, let's try this to decide whether $1642 \mid 136299$. Start finding multiples of 1642:

$$1642 \cdot 2 = 3284 \quad 1642 \cdot 3 = 49C26 \quad 1642 \cdot 4 = 6568 \cdots.$$

All of these are well less than 136299. I suppose we can jump ahead a bit:

$$1642 \cdot 50 = 82100 \quad 1642 \cdot 80 = 131360 \quad 1642 \cdot 85 = 139570.$$

Ah, so we need to look somewhere between 80 and 85. Try 83:

$$1642 \cdot 83 = 136286.$$

Is this the best we can do? How far are we from our desired 136299? If we subtract, we get $136299 - 136286 = 13$. So we know we cannot go up to 84, that will be too much. In other words, we have found that

$$136299 = 83 \cdot 1642 + 13.$$

Since $13 < 1642$, we can now safely say that $1642 \nmid 136299$.

It turns out that the process we went through above can be repeated for any pair of numbers. We can always write the number a as some multiple of the number b plus some remainder. We know this because we know about division with remainder from elementary school. This is just a way of saying it using multiplication. Due to the procedural nature that can be used to find the remainder, this fact is usually called the division algorithm:

The Division Algorithm

Given any two integers a and b, we can always find an integer q such that

$$a = qb + r$$

where r is an integer satisfying $0 \le r < |b|$

The idea is that we can always take a large enough multiple of b so that the remainder r is as small as possible. We do allow the possibility of r = 0, in which case we have b | a.

Remainder Classes

The division algorithm tells us that there are only b possible remainders when dividing by b. If we fix this divisor, we can group integers by the remainder. Each group is called a remainder class modulo b (or sometimes residue class).

Example 6.2.2

Describe the remainder classes modulo 5.

Solution

We want to classify numbers by what their remainder would be when divided by 5. From the division algorithm, we know there will be exactly 5 remainder classes, because there are only 5 choices for what r could be ($0 \le r < 5$).

First consider r = 0. Here we are looking for all the numbers divisible by 5 since a = 5q + 0. In other words, the multiples of 5.

We get the infinite set

$$\{\ldots, -15, -10, -5, 0, 5, 10, 15, 20, \ldots\}.$$

Notice we also include negative integers.

Next consider r = 1. Which integers, when divided by 5, have remainder 1? Well, certainly 1, does, as does 6, and 11. Negatives?

Here we must be careful: −6 does NOT have remainder 1. We can write $-6 = -2 \cdot 5 + 4$ or $-6 = -1 \cdot 5 - 1$, but only one of these is a "correct" instance of the division algorithm: r = 4 since we need r to be non-negative. So in fact, to get r = 1, we would have −4, or −9, etc. Thus we get the remainder class

$$\{\ldots, -14, -9, -4, 1, 6, 11, 16, 21, \ldots\}.$$

There are three more to go. The remainder classes for 2, 3, and 4 are, respectively

$$\{\ldots, -13, -8, -3, 2, 7, 12, 17, 22, \ldots\}$$

$$\{\ldots, -12, -7, -2, 3, 8, 13, 18, 23, \ldots\}$$

$$\{\ldots, -11, -6, -1, 4, 9, 14, 19, 24, \ldots\}.$$

Note that in the example above, every integer is in exactly one remainder class. The technical way to say this is that the remainder classes modulo b form a partition of the integers.1 The most important fact about partitions, is that it is possible to define an equivalence relation from a partition: this is a relationship between pairs of numbers which acts in all the important ways like the "equals" relationship.2

All fun technical language aside, the idea is really simple. If two numbers belong to the same remainder class, then in some way, they are the same. That is, they are the same up to division by b. In the case where b = 5 above, the numbers 8 and 23, while not the same number, are the same when it comes to dividing by 5, because both have remainder 3.

It matters what the divisor is: 8 and 23 are the same up to division by 5, but not up to division by 7, since 8 has remainder of 1 when divided by 7 while 23 has a remainder of 2.

With all this in mind, let's introduce some notation. We want to say that 8 and 23 are basically the same, even though they are not equal. It would be wrong to say $8 \equiv 23$. Instead, we write 8 23. But this is not always true. It works if we are thinking division by 5, so we need to denote that somehow. What we will actually write is this:

$$8 \equiv 23 \pmod 5$$

which is read, "8 is congruent to 23 modulo 5" (or just "mod 5"). Of course then we could observe that

$$8 \equiv 23 \pmod 7$$

Congruence Modulo n.

We say a is congruent to b modulo n, and write,

$$a \equiv b \pmod n$$

provided a and b have the same remainder when divided by n. In other words, provided a and b belong to the same remainder class modulo n.

Many books define congruence modulo n slightly differently. They say that a b mod n if and only if n a b. In other words, two numbers are congruent modulo n, if their difference is a multiple of n. So which definition is correct? Turns out, it doesn't matter: they are equivalent.

To see why, consider two numbers a and b which are congruent modulo n. Then a and b have the same remainder when divided by n. We have

$$a = q1n + r \quad b = q2n + r.$$

Here the two r's really are the same. Consider what we get when we take the difference of a and b:

$$a - b = q_1 n + r - (q_2 n + r) = q_1 n - q_2 n = (q_1 - q_2)n.$$

So a– b is a multiple of n, or equivalently, n| a– b.

On the other hand, if we assume first that n |a– b, so a– b = kn, then consider what happens if we divide each term by n. Dividing a by n will leave some remainder, as will dividing b by n. However, dividing kn by n will leave 0 remainder. So the remainders on the left-hand side must cancel out. That is, the remainders must be the same.

Thus we have:

- Congruence and Divisibility.
- For any integers a, b, and n, we have
- $a \equiv b \pmod{n}$ if and only if $n \mid a - b$.

It will also be useful to switch back and forth between congruences and regular equations. The above fact helps with this. We know that $a \equiv b \pmod{n}$ if and only if n |a– b, if and only if a– b = kn for some integer k. Rearranging that equation, we got a = b + kn. In other words, if a and b are congruent modulo n, then a is b more than some multiple of n. This conf orms with our earlier observation that all the numbers in a particular remainder class are the same amount larger than the multiples of n.

Congruence and Equality.

For any integers a, b, and n, we have

$a \equiv b \pmod{n}$ if and only if $a = b+kn$ for some integer k.

Properties of Congruence

We said earlier that congruence modulo n behaves, in many important ways, the same way equality does. Specifically, we could prove that congruence modulo n is an equivalence relation, which would require checking the following three facts:

Congruence Modulo n is an Equivalence Relation.

Given any integers a, b, and c, and any positive integer n, the following hold:

- $a \equiv a \pmod{n}$.
- If $a \equiv b \pmod{n}$ then $b \equiv a \pmod{n}$.
- If $a \equiv b \pmod{n}$ and $b \equiv c \pmod{n}$, then $a \equiv c \pmod{n}$.

In other words, congruence modulo n is reflexive, symmetric, and transitive, so is an equivalence relation.

You should take a minute to convince yourself that each of the properties above actually hold of congruence. Try explaining each using both the remainder and divisibility definitions.

Next, consider how congruence behaves when doing basic arithmetic. We already know that if you subtract two congruent numbers, the result will be congruent to 0 (be a multiple of n). What if we add something congruent to 1 to something congruent to 2? Will we get something congruent to 3?

Congruence and Arithmetic

Suppose $a \equiv b \pmod{n}$ and $c \equiv d \pmod{n}$. Then the following hold:

$a + c \equiv b + d \pmod{n}$.

$a - c \equiv b - d \pmod{n}$.

$ac \equiv bd \pmod{n}$.

The above facts might be written a little strangely, but the idea is simple. If we have a true congruence, and we add the same thing to both sides, the result is still a true congruence. This sounds like we are saying:

$$\text{If } a \equiv b \pmod{n} \text{ then } a + c \equiv b + c \pmod{n}.$$

Of course this is true as well, it is the special case where $c = d$. But what we have works in more generality. Think of congruence as being "basically equal." If we have two numbers which are basically equal, and we add basically the same thing to both sides, the result will be basically equal.

This seems reasonable. Is it really true? Let's prove the first fact:

Proof. Suppose $a \equiv b \bmod n$ and $c \equiv d \bmod n$. That means $a = b + kn$ and $c = d + jn$ for integers k and j. Add these equations:

$$a + c = b + d + kn + jn.$$

But $kn + jn = (k + j)n$, which is just a multiple of n. So $a + c = b + d + (j + k)n$, or in other words, $a + c \equiv b + d \pmod{n}$

The other two facts can be proved in a similar way.

One of the important consequences of these facts about congruences, is that we can basically replace any number in a congruence with any other number it is congruent to. Here are some examples to see how (and why) that works:

Example 6.2.3

Find the remainder of 3491 divided by 9.

Solution

We could do long division, but there is another way. We want to find x such that x 3491 mod 9 . Now $3491 = 3000 + 400 + 90 + 1$. Of course 90 0 mod 9 , so we can replace the 90 in the sum with 0. Why is this okay? We are actually subtracting the "same" thing from both sides:

$x \equiv 3000 + 400 + 90 + 1 \pmod 9$

$-0 \equiv 90 \pmod 9$

$x \equiv 3000 + 400 + 0 + 1 \pmod 9$.

Next, note that 400 = 4 100, and 100 1 mod 9 (since 9 99). So we can in fact replace the 400 with simply a 4. Again, we are appealing to our claim that we can replace congruent elements, but we are really appealing to property 3 about the arithmetic of congruence: we know $100 \equiv 1 \pmod 9$, so if we multiply both sides by 4, we get $400 \equiv 4 \pmod 9$.

Similarly, we can replace 3000 with 3, since $1000 = 1 + 999 \equiv 1 \pmod 9$. So our original congruence becomes

$x \equiv 3 + 4 + 0 + 1 \pmod 9$

$x \equiv 8 \pmod 9$.

Therefore 3491 divided by 9 has remainder 8.

The above example should convince you that the well known divisibility test for 9 is true; the sum of the digits of a number is divisible by 9 if and only if the original number is divisible by 9. In fact, we now know something more: any number is congruent to the sum of its digits, modulo 9.3

Let's try another:

Example 6.2.4

Find the remainder when 3123 is divided by 7.

Solution

Of course, we are working with congruence because we want to find the smallest positive x such that $x \equiv 3123 \pmod 7$.

Now first write 3123 = (33)41. We have:

$3123 = 2741 \equiv 641 \pmod 7$,

since $27 \equiv 6 \pmod 7$. Notice further that 62 = 36 is congruent to 1 modulo 7. Thus we can simplify further:

$641 = 6 \cdot (62)20 \equiv 6 \cdot 120 \pmod 7$.

But 120 = 1, so we are done:

$3123 \equiv 6 \pmod 7$.

In the above example, we are using the fact that if $a \equiv b \bmod n$, then $ap \equiv bp \bmod n$. This is just applying property 3 a bunch of times.

So far we have seen how to add, subtract and multiply with congruences.

What about division? There is a reason we have waited to discuss it. It turns out that we cannot simply divide. In other words, even if ad ≡ bd (mod n), we do not know that a ≡ b (mod n). Consider, for example:

$18 \equiv 42 \pmod{8}$.

This is true. Now 18 and 42 are both divisible by 6. However,

$3 \equiv 7 \pmod{8}$.

While this doesn't work, note that $3 \equiv 7 \bmod 4$. We cannot divide 8 by 6, but we can divide 8 by the greatest common factor of 8 and 6. Will this always happen?

Suppose ad ≡ bd mod n. In other words, we have ad = bd + kn for some integer k. Of course ad is divisible by d, as is bd. So kn must also be divisible by d. Now if n and d have no common factors (other than 1), then we must have d|k. But in general, if we try to divide kn by d, we don't know that we will get an integer multiple of n. Some of the n might get divided as well. To be safe, let's divide as much of n as we can. Take the largest factor of both d and n, and cancel that out from n. The rest of the factors of d will come from k, no problem.

We will call the largest factor of both d and n the gcd (d, n), for greatest common divisor. In our example above, gcd (6, 8) = 2 since the greatest divisor common to 6 and 8 is 2.

Congruence and Division

Suppose ad ≡ bd (mod n). Then $\left(\bmod \frac{n}{\gcd(d,n)}\right)$.

If d and n have no common factors then gcd(d, n) = 1, so a ≡ b (mod n).

Example 6.2.5

Simplify the following congruences using division: (a) 24 ≡ 39 (mod 5) and (b) 24 ≡ 39 (mod 15).

Solution

(a) Both 24 and 39 are divisible by 3, and 3 and 5 have no common factors, so we get

$$8 \equiv 13 \pmod{5}.$$

(b) Again, we can divide by 3. However, doing so blindly gives us 8 ≡ 13 (mod 15) which is no longer true. Instead, we must also divide the modulus 15 by the greatest common factor of 3 and 15, which is 3. Again we get

$$8 \equiv 13 \pmod{5}.$$

Solving Congruences

Now that we have some algebraic rules to govern congruence relations, we can attempt to solve for an unknown in a congruence. For example, is there a value of x that satisfies,

$$3x + 2 \equiv 4 \pmod{5},$$

and if so, what is it?

In this example, since the modulus is small, we could simply try every possible value for x. There are really only 5 to consider, since any integer that satisfied the congruence could be replaced with any other integer it was congruent to modulo 5. Here, when $x = 4$ we get $3x + 2 = 14$ which is indeed congruent to 4 modulo 5. This means that $x = 9$ and $x = 14$ and $x = 19$ and so on will each also be a solution because as we saw above, replacing any number in a congruence with a congruent number does not change the truth of the congruence.

So in this example, simply compute $3x + 2$ for values of $x \in \{0, 1, 2, 3, 4\}$. This gives 2, 5, 8, 11, and 14 respectively, for which only 14 is congruent to 4.

Let's also see how you could solve this using our rules for the algebra of congruences. Such an approach would be much simpler than the trial and error tactic if the modulus was larger. First, we know we can subtract 2 from both sides:

$$3x \equiv 2 \pmod{5}.$$

Then to divide both sides by 3, we first add 0 to both sides. Of course, on the right-hand side, we want that 0 to be a 10 (yes, 10 really is 0 since they are congruent modulo 5). This gives,

$$3x \equiv 12 \pmod{5}.$$

Now divide both sides by 3. Since $\gcd(3, 5) = 1$, we do not need to change the modulus:

$$x \equiv 4 \pmod{5}.$$

Notice that this in fact gives the general solution: not only can $x = 4$, but x can be any number which is congruent to 4. We can leave it like this, or write "$x = 4 + 5k$ for any integer k."

Example 6.2.6

Solve the following congruences for x.

$$7x \equiv 12 \pmod{13}.$$

$$84x - 38 \equiv 79 \pmod{15}.$$

$$20x \equiv 23 \pmod{14}.$$

Solution

All we need to do here is divide both sides by 7. We add 13 to the right-hand side repeatedly until we get a multiple of 7 (adding 13 is the same as adding 0, so this is legal). We get 25, 38, 51, 64, 77 – got it. So we have:

$$7x \equiv 12 \pmod{13}$$
$$7x \equiv 77 \pmod{13}$$
$$x \equiv 11 \pmod{13}.$$

Here, since we have numbers larger than the modulus, we can reduce them prior to applying any algebra. We have $84 \equiv 9$, $38 \equiv 8$ and $79 \equiv 4$. Thus,

$$84x - 38 \equiv 79 \pmod{15}$$
$$9x - 8 \equiv 4 \pmod{15}$$
$$9x \equiv 12 \pmod{15}$$
$$9x \equiv 72 \pmod{15}.$$

We got the 72 by adding $0 \equiv 60 \pmod{15}$ to both sides of the congruence. Now divide both sides by 9. However, since $\gcd(9, 15) = 3$, we must divide the modulus by 3 as well:

$$x \equiv 8 \pmod{5}.$$

So the solutions are those values which are congruent to 8, or equivalently 3, modulo 5. This means that in some sense there are 3 solutions modulo 15: 3, 8, and 13. We can write the solution:

$$x \equiv 3 \pmod{15};\ x \equiv 8 \pmod{15};\ x \equiv 13 \pmod{15}.$$

First, reduce modulo 14:

$$20x \equiv 23 \pmod{14}$$
$$6x \equiv 9 \pmod{14}.$$

We could now divide both sides by 3, or try to increase 9 by a multiple of 14 to get a multiple of 6. If we divide by 3, we get,

$$2x \equiv 3 \pmod{14}.$$

Now try adding multiples of 14 to 3, in hopes of getting a number we can divide by 2. This will not work! Every time we add 14 to the right side, the result will still be odd. We will never get an even number, so we will never be able to divide by 2. Thus there are no solutions to the congruence.

The last congruence above illustrates the way in which congruences might not have solutions. We could have seen this immediately in fact. Look at the original congruence:

$$20x \equiv 23 \pmod{14}.$$

If we write this as an equation, we get

$$20x = 23 + 14k,$$

or equivalently $20x - 14k = 23$. We can easily see there will be no solution to this equation in integers. The left-hand side will always be even, but the right-hand side is odd. A similar problem would occur if the right-hand side was divisible by any number the left-hand side was not.

So in general, given the congruence

$$ax \equiv b \pmod{n},$$

Congruences with no solutions.

If $\gcd(a, n) \nmid b$, then $ax \equiv b \pmod{n}$ has no solutions.

Solving Linear Diophantine Equations

Discrete math deals with whole numbers of things. So when we want to solve equations, we usually are looking for integer solutions. Equations which are intended to only have integer solutions were first studied by in the third century by the Greek mathematician Diophantus of Alexandria, and as such are called Diophantine equations. Probably the most famous example of a Diophantine equation is $a2 + b2 = c2$. The integer solutions to this equation are called Pythagorean triples. In general, solving Diophantine equations is hard (in fact, there is provably no general algorithm for deciding whether a Diophantine equation has a solution, a result known as Matiyasevich's Theorem). We will restrict our focus to linear Diophantine equations, which are considerably easier to work with.

Diophantine Equations

An equation in two or more variables is called a Diophantine equation if only integers solutions are of interest. A linear Diophantine equation takes the form $a1x1 + a2xx + \cdots + anxn = b$ for constants a1, . . . , an , b.

A solution to a Diophantine equation is a solution to the equation consisting only of integers.

We have the tools we need to solve linear Diophantine equations. We will consider, as a main example, the equation

$$51x + 87y = 123.$$

The general strategy will be to convert the equation to a congruence, then solve that congruence.4 Let's work this particular example to see how this might go.

First, check if perhaps there are no solutions because a divisor of 51 and 87 is not a divisor of 123. Really, we just need to check whether gcd (51, 87) |123. This greatest common divisor is 3, and yes 3|123. At this point, we might as well factor out this greatest common divisor. So instead, we will solve:

$$17x + 29y = 41.$$

Now observe that if there are going to be solutions, then for those values of x and y, the two sides of the equation must have the same remainder as each other, no matter what we divide by. In particular, if we divide both sides by 17, we must get the same remainder. Thus we can safely write

$$17x + 29y \equiv 41 \pmod{17}.$$

We choose 17 because 17x will have remainder 0. This will allow us to reduce the congruence to just one variable. We could have also moved to a congruence modulo 29, although there is usually a good reason to select the smaller choice, as this will allow us to reduce the other coefficient. In our case, we reduce the congruence as follows:

$$17x + 29y \equiv 41 \pmod{17}$$
$$0x + 12y \equiv 7 \pmod{17}$$
$$12y \equiv 24 \pmod{17}$$
$$y \equiv 2 \pmod{17}.$$

Now at this point we know $y = 2 + 17k$ will work for any integer k. If we haven't made a mistake, we should be able to plug this back into our original Diophantine equation to find x:

$$17x + 29(2 + 17k) = 41$$
$$17x = -17 - 29 \cdot 17k$$
$$x = -1 - 29k.$$

We have now found all solutions to the Diophantine equation. For each k, $x =- 1 -29k$ and $y = 2 + 17k$ will satisfy the equation. We could check this for a few cases. If $k = 0$, the solution is $(-1, 2)$, and yes, $-17 + 2.29 = 41$. If $k = 3$, the solution is $(-88, 53)$. If $k = -2$, we get $(57, 32)$.

To summarize this process, to solve $ax + by = c$, we,

Divide both sides of the equation by gcd (a, b) (if this does not leave the right-hand side as an integer, there are no solutions). Let's assume that $ax + by = c$ has already been reduced in this way.

Pick the smaller of a and b (here, assume it is b), and convert to a congruence modulo b:

$$ax + by \equiv c \pmod{b}.$$

This will reduce to a congruence with one variable, x:

$$ax \equiv c \pmod{b}.$$

Solve the congruence as we did in the previous section. Write your solution as an equation, such as,

$$x = n + kb.$$

Plug this into the original Diophantine equation, and solve for y.

If we want to know solutions in a particular range (for example, $0 \le x, y \le 20$), pick different values of k until you have all required solutions.

Here is another example:

Example 6.2.7

How can you make $6.37 using just 5-cent and 8-cent stamps? What is the smallest and largest number of stamps you could use?

Solution

First, we need a Diophantine equation. We will work in numbers of cents. Let x be the number of 5-cent stamps, and y be the number of 8-cent stamps. We have:

$$5x + 8y = 637.$$

Convert to a congruence and solve:

$$8y \equiv 637 \pmod 5$$
$$3y \equiv 2 \pmod 5$$
$$3y \equiv 12 \pmod 5$$
$$y \equiv 4 \pmod 5.$$

Thus $y = 4 + 5k$. Then $5x + 8(4 + 5k) = 637$, so $x = 121 - 8k$.

This says that one way to make $6.37 is to take 121 of the 5-cent stamps and 4 of the 8-cent stamps. To find the smallest and largest number of stamps, try different values of k.

k	(x, y)	Stamps
-1	(129, -1)	not possible
0	(121, 4)	125
1	(113, 9)	122
2	(105, 13)	119
. . .	. . .	. . .

This is no surprise. Having the most stamps means we have as many 5-cent stamps as possible, and to get the smallest number of stamps would require have the least number of 5-cent stamps. To minimize the number of 5-cent stamps, we want to pick k so that 121– 8k is as small as possible (but still positive). When k = 15, we have x = 1 and y = 79.

Therefore, to make \$6.37, you can us as few as 80 stamps (1 5-cent stamp and 79 8-cent stamps) or as many as 125 stamps (121 5-cent stamps and 4 8-cent stamps).

Using this method, as long as you can solve linear congruences in one variable, you can solve linear Diophantine equations of two variables. There are times though that solving the linear congruence is a lot of work. For example, suppose you need to solve,

$$13x \equiv 6 \pmod{51}.$$

You could keep adding 51 to the right side until you get a multiple of 13: You would get 57, 108, 159, 210, 261, 312, and 312 is the first of these that is divisible by 13. This works, but is really too much work. Instead we could convert back to a Diophantine equation:

$$13x = 6 + 51k.$$

Now solve this like we have in this section. Write it as a congruence modulo 13:

$$0 \equiv 6 + 51k \pmod{13}$$

$$-12k \equiv 6 \pmod{13}$$

$$2k \equiv -1 \pmod{13}$$

$$2k \equiv 12 \pmod{13}$$

$$k \equiv 6 \pmod{13}.$$

so k = 6 + 13 j. Now go back and figure out x:

$$13x = 6 + 51(6 + 13j)$$

$$x = 24 + 51j.$$

Of course you could do this switching back and forth between congruences and Diophantine equations as many times as you like. If you only used this technique, you would essentially replicate the Euclidean algorithm, a more standard way to solve Diophantine equations.

EXERCISES

1. **Suppose a, b, and c are integers. Prove that if a|b, then a |bc.**
2. **Suppose a, b, and c are integers. Prove that if a | b and a | c then a | b + c and a| b - c.**
3. **Write out the remainder classes for n = 4.**
4. **What is the largest n such that 16 and 25 are in the same remainder class modulo n? Write out the remainder class they both belong to and give an example of a number more than 100 in that class.**

5. **Let a, b, c, and n be integers. Prove that if a ≡ b (mod n) and c ≡ d (mod n), then a - c ≡ b - d (mod n).**

6. **Find the remainder of 3456 when divided by**

   (a) 2. (b) 5.

   (c) 7. (d) 9.

7. **Repeat the previous exercise, this time dividing 22019.**

8. **Determine which of the following congruences have solutions, and find any solutions (between 0 and the modulus) by trial and error.**

   (a) $4x \equiv 5 \pmod 6$.

   (b) $6x \equiv 3 \pmod 9$.

   (c) $x2 \equiv 2 \pmod 4$.

9. **Determine which of the following congruences have solutions, and find any solutions (between 0 and the modulus) by trial and error.**

   (a) $4x \equiv 5 \pmod 7$.

   (b) $6x \equiv 4 \pmod 9$.

   (c) $x2 \equiv 2 \pmod 7$.

10. **Solve the following congruence $5x + 8 \equiv 11 \pmod{22}$. That is, describe the general solution.**

11. **Solve the congruence: $6x \equiv 4 \pmod{10}$.**

12. **Solve the congruence: $4x \equiv 24 \pmod{30}$.**

13. **Solve the congruence: $341x \equiv 2941 \pmod 9$.**

14. **I'm thinking of a number. If you multiply my number by 7, add 5, and divide the result by 11, you will be left with a remainder of 2. What remainder would you get if you divided my original number by 11?**

15. **Solve the following linear Diophantine equation, using modular arithmetic (describe the general solutions).**

$$6x + 10y = 32.$$

16. **Solve the following linear Diophantine equation, using modular arithmetic (describe the general solutions).**

$$17x + 8y = 31.$$

17. **Solve the following linear Diophantine equation, using modular arithmetic (describe the general solutions).**

$$35x + 47y = 1.$$

18. **You have a 13 oz. bottle and a 20 oz. bottle, with which you wish to measure exactly 2 oz. However, you have a limited supply of water. If any water enters either bottle and then gets dumped out, it is gone forever. What is the least amount of water you can start with and still complete the task?**

INDEX

F

G

H

I

K

L

M

N

O

P

Q

R

S